T0234012

SpringerBriefs in Applied Sciences and Technology

Safety Management

Series Editors

Eric Marsden, FonCSI, Toulouse, France

Caroline Kamaté, FonCSI, Toulouse, France

François Daniellou, FonCSI, Toulouse, France

The SpringerBriefs in Safety Management present cutting-edge research results on the management of technological risks and decision-making in high-stakes settings.

Decision-making in high-hazard environments is often affected by uncertainty and ambiguity; it is characterized by trade-offs between multiple, competing objectives. Managers and regulators need conceptual tools to help them develop risk management strategies, establish appropriate compromises and justify their decisions in such ambiguous settings. This series weaves together insights from multiple scientific disciplines that shed light on these problems, including organization studies, psychology, sociology, economics, law and engineering. It explores novel topics related to safety management, anticipating operational challenges in high-hazard industries and the societal concerns associated with these activities.

These publications are by and for academics and practitioners (industry, regulators) in safety management and risk research. Relevant industry sectors include nuclear, offshore oil and gas, chemicals processing, aviation, railways, construction and healthcare. Some emphasis is placed on explaining concepts to a non-specialized audience, and the shorter format ensures a concentrated approach to the topics treated.

The SpringerBriefs in Safety Management series is coordinated by the Foundation for an Industrial Safety Culture (FonCSI), a public-interest research foundation based in Toulouse, France. The FonCSI funds research on industrial safety and the management of technological risks, identifies and highlights new ideas and innovative practices, and disseminates research results to all interested parties.

For more information: https://www.foncsi.org/.

FONCSI
Fondation pour une culture
de sécurité industrielle

More information about this subseries at https://link.springer.com/bookseries/15119

Jan Hayes · Stéphanie Tillement
Editors

Contracting and Safety

Exploring Outsourcing Practices
in High-Hazard Industries

 Springer

Editors
Jan Hayes
School of Property, Construction
and Project Management
RMIT University
Melbourne, VIC, Australia

Stéphanie Tillement
Département Sciences Sociales et de
Gestion
IMT Atlantique
Nantes, France

Fondation pour une Culture de Securite Industrielle

ISSN 2191-530X   ISSN 2191-5318 (electronic)
SpringerBriefs in Applied Sciences and Technology
ISSN 2520-8004   ISSN 2520-8012 (electronic)
SpringerBriefs in Safety Management
ISBN 978-3-030-89791-8   ISBN 978-3-030-89792-5 (eBook)
https://doi.org/10.1007/978-3-030-89792-5

This Springer imprint is published by the registered company Springer Nature Switzerland AG
The registered company address is: Gewerbestrasse 11, 6330 Cham, Switzerland

Foreword

One of the distinctive features of humankind is its extraordinary capacity to organise to achieve its purposes—from the building of the pyramids to the building of a modern jet aircraft, from fighting a war to travelling to the moon. Such activities are quite beyond any individual or small group of individuals and require the coordination of thousands, if not millions, of human beings. The organisation, in all its forms, is thus a central concern of sociology, starting with Max Weber and his claim that bureaucracy is technically superior to any other form of organisation.

More specifically, sociology has long been interested in the relationship between organisational characteristics and safety. One of the earliest examples is Alvin Gouldner's 1954 study of bureaucratisation and safety in a gypsum plant. Better known nowadays is Charles Perrow's 1984 study of *normal accidents*, which connects accidents to organisational complexity and coupling. Another landmark in this terrain is Jim Reason's book, *Organisational Accidents*. My own work, and that of my colleague Gilsa Monteiro, has sought to show that in hazardous industries the decentralisation of responsibility for the management of catastrophic risk is a contributing factor to major accidents.

Much of the aforementioned research has focused on single organisations. But the growth of subcontracting or outsourcing since the late twentieth century has changed the way organisations operate. There are now often networks or chains of contractors reaching from the principal organisation down to microbusinesses, sometimes single operators. Students of Industrial Relations have been calling attention to this for many years, highlighting the exploitative relations that exist at the lower end of such chains and coining the term *precarious employment* to describe the circumstances of such workers. But the connection between this contemporary organisational phenomenon and safety is less well studied. This book is a major contribution to filling the gap.

What is so refreshing about the book is that it views safety as a system property, not just a question of individual behaviour. The contributors focus on the daily practices and interactions between the principal and the contractor, but they relate this directly back to the organisational system in which these people are operating. Understanding the deep roots of safe and unsafe outcomes in this way will help identify changes that can be made to improve workplace safety.

The authors are internationally recognised scholars in their fields, and I am very pleased to be able to commend this book wholeheartedly to the reader.

Andrew Hopkins
Emeritus Professor of Sociology
Australian National University
Canberra, Australia

Preface

Safety researchers have a strong interest in organisations because we understand that the way in which work is organised impacts on safety outcomes. Despite this, when it comes to outsourcing and contracting, changes in the way in which work is structured have outpaced organisational safety research. Keeping a complex sociotechnical system safe is a long-term endeavour which poses particular challenges given the potentially short-term and transnational nature of contract relationships and yet these two have been little studied in parallel.

Investigating this gap was the driving force behind two sessions convened at the Society for Social Studies of Science (4S) Annual Conference in Sydney, Australia, in August 2018. These sessions on *The Impact of Outsourcing and Contracting on Accident Prevention in Complex Sociotechnical Systems* brought together interested researchers from Europe, Asia and Australia with an aim to better understand how safety (and reliability) can be achieved in these distributed (and often unique) settings. These conversations ultimately resulted in the chapters in this book which provide a starting point to understanding how to ensure that contracting benefits all aspects of organisational performance instead of becoming a trade-off between cost and safety.

The co-editors, Jan Hayes and Stéphanie Tillement, are deeply grateful to the ANR/Investissements d'Avenir AGORAS[1] programme for funding this publication and to the RESOH Chair[2] and Foundation for an Industrial Safety Culture (FonCSI)[3] for support for this research initiative.

---

[1] AGORAS (*Amélioration de la Gouvernance des organisations et des Réseaux d'Acteurs pour la Sûreté nucléaire*—Improving the governance of organisations and actors' networks for nuclear safety): https://web.imt-atlantique.fr/x-ssg/projetagoras.

[2] RESOH (Research on Safety Organisation Humans): https://www.imt-atlantique.fr/resoh.

[3] FonCSI: https://www.foncsi.org/.

We would also like to thank Ms. Janice Wong for her excellent support in preparation of the manuscript.

Melbourne Australia   Jan Hayes
Nantes, France   Stéphanie Tillement
May 2021

Contents

1   Outsourcing and Safety—An Introduction 1
    Jan Hayes and Stéphanie Tillement

2   Contracting and Safety: Lessons from Observing
    an Outsourcing Process "in the Making" 9
    Stéphanie Tillement and Geoffrey Leuridan

3   Workload Planning Management of Maintenance Activities
    in Nuclear Power Plants: Compensation Mechanisms
    at the Contractor Interface 19
    Nicolas Dechy and Alexandre Largier

4   Inter-organisational Collaboration for the Safety of Railway
    Vehicles: A Japanese Case 29
    Takuji Hara

5   Engaged Scholarship for Exploring Applicability of Relational
    Contracting to Nuclear Industry Projects 39
    Nadezhda Gotcheva, Kirsi Aaltonen, Pertti Lahdenperä,
    and Soili Nysten-Haarala

6   Contracting Qualities that Challenge Reliability: A Case
    of the Utility Sector 49
    Léon L. olde Scholtenhuis

7   Managing Workplace Safety in the Temporary
    Organisation—Theoretical and Practical Challenges
    Associated with Large Construction Projects 59
    Heidi Helledal Griegel and Kenneth Pettersen Gould

8   When the Project Ends and Operations Begin: Ensuring
    Safety During Commissioning Through Boundary Work 69
    Anne Russel and Stéphanie Tillement

**9   Outsourcing Risk Governance: Using Consultants to Deliver
      Regulatory Functions**   79
      Jan Hayes, Lynne Chester, and Dolruedee Kramnaimuang King

**10   Outsourcing in Theory and Practice: Insights from Nuclear
      Risk Governance**   89
      Jérémy Eydieux

**11   Outsourced Enforcement: Improving the Public
      Accountability of Building Inspectors**   99
      Nader Naderpajouh, Rita Peihua Zhang, and Jan Hayes

12   Implications for Safe Outsourcing   109
      Stéphanie Tillement and Jan Hayes

Chapter 1
Outsourcing and Safety—An Introduction

Jan Hayes and Stéphanie Tillement

Abstract Organisational safety theory and research has not kept pace with changing trends in the way work is organised. Outsourcing has become very common and is known to be problematic for workplace safety. This book aims to extend consideration of the link between safety and outsourcing to system safety.

Keywords Outsourcing · Contracting · Safety · System safety · Work practices

1.1 Why Outsourcing and Safety?

In business parlance, outsourcing means contracting out a portion of a company's business activity to an external party, rather than doing it in-house. The contracted activity may be provision of a service or production of goods. This business practice has become ubiquitous since the latter part of the twentieth century with many companies choosing to conduct their activities using workers who are not directly employed by them. This is done to maximise profits using external resources to manage peaks in workload (which may be activity-related, seasonal or unplanned) and/or to provide specialist skills that are seen as no longer core to the organisation itself [1]. Cost savings arise as activities move from being subject to internal governance and are placed under the governance of the market.

Outsourcing has become so ubiquitous that questions have been asked as to whether its use has gone too far [5] and yet a recent study demonstrated the enduring popularity of outsourcing finding that the primary reason for its adoption remains cost reduction [4]. Coincidentally, many organisations are using more complex supply networks and yet putting less effort into supplier management to the point where

J. Hayes (✉)
RMIT University, Melbourne, Australia
e-mail: jan.hayes2@rmit.edu.au

S. Tillement
IMT Atlantique, LEMNA, Nantes, France
e-mail: stephanie.tillement@imt-atlantique.fr

© The Author(s) 2022
J. Hayes and S. Tillement (eds.), *Contracting and Safety*,
SpringerBriefs in Safety Management,
https://doi.org/10.1007/978-3-030-89792-5_1

'some service providers have characterised their client vendor management functions as "chaotic" and "purely cost-focused".' [4, p. 12].

We use the terms outsourcing and contracting interchangeably to represent this way of organising work. In fact, often there are chains of contractors, each providing flexibility in the form of resources and/or expertise for the organisation above it in the chain. This potentially extends all the way to microbusinesses made up of only one person (or perhaps a family group). From the perspective of workers, this on-demand work pushes them into the gig economy where workers no longer hold many of the hard-won rights to paid sick leave and some level of job security that are afforded to employees. Online technologies have made real-time matching of available workers and those seeking services a reality [9]. This may hold the promise of boosting the economy, but precarious work of this kind is known to have an adverse impact on worker health and safety [10, 11, 13, 16].

Safety and outsourcing might be seen as an issue primarily for the services sector including hospitality, ridesharing services or personal care service workers, but such segmentation of the working environment has safety implications for high-hazard industry too. Major projects have always been born out of temporary organisations that bring together organisations with varying expertise allowing the ultimate facility owner/operator to 'do what they do best' and outsource everything else. Since the 1980s, outsourcing during the operational phase of hazardous facilities has become common, particularly with regard to maintenance activities [6]. This brings outsourcing into industries where system safety–the prevention of major disasters with the potential to cause significant numbers of public deaths and destruction of property–is a significant concern, in addition to the day-to-day safety and health of workers.

In contrast to these flexible but sometimes complex networks of organisations, the safety literature that focuses on organisational accident causation of major disasters tends to assume that, at least operationally, work is conducted by one organisation. This is true of the well-known Swiss cheese model [15] and also for research into high-reliability organisations. In her retrospective interview, Karlene Roberts highlighted the issue of organisational 'interstices' emphasising that more research was needed to understand how errors should be managed at 'places where parts of organisations come together' [2, p. 95]. In 2021, interstices worthy of study go well beyond interdepartmental boundaries and other such internal organisational interfaces to include boundaries between different organisations that work closely together with a common goal.

The way in which work is organised has also been seen as a causal factor in several major disasters. The official commission of inquiry into the 2001 Toulouse explosion at the AZF chemical factory found that complex chains of contractors and subcontractors were a critical factor in the accident [3, 14]. As result, the inquiry recommended that multi-tier contractor arrangements be banned on major hazard sites. The causes of the Deepwater Horizon blowout that so devastated the coastal environment around the Gulf of Mexico in 2010 were also linked to the way in which offshore exploration drilling is organised using consortia of specialist contractors [7].

Consideration of the impact of organisational reporting lines and hierarchies on the potential for major disaster has led to recent appeals for a greater focus on organisational structure and its impact on accident causation [8]. To Hopkins, the notion of structure refers mostly to the degree to which the way of managing major hazards is centralised or decentralised. It also designates the communication and reporting lines (notably with respect to safety concerns), the structure of incentives and the company priorities that are set by CEOs, boards and remuneration systems. Culture is defined as 'the way we do things around here' [8, p. 29]. This simple definition has several implications. Firstly, it means that culture is necessarily a characteristic of a group, whatever its size, and so inherently associated with collective practices. Secondly, Hopkins insists on the normative dimension of this definition, the 'way of doing' implicitly being 'the right, or appropriate, or accepted way to do things' [8, p. 30]. Thus, the emphasis on practice 'does not exclude the importance of norms and values.' Following this, 'the culture of safety is a way of doing things that emphasises safety' [8, p. 28]. Since collective practices largely depend on organisational characteristics, structure is a key variable in improving safety and preventing major hazards. In this view, a poor culture is not so much a cause of accidents but a consequence of a pathological organisational structure. If structure creates safety culture as Hopkins contends, that gives us even more reason to study the impact of outsourcing and the new structures it imposes on work. With these theoretical considerations in mind, we need research methods to study the impact of outsourcing on work practices on the ground.

1.2 Research Approach

Contrary to most existing studies of outsourcing that typically adopt a rather quantitative and macro-approach (based on analysis of incident data bases for example), the vast majority of the chapters in this book use a qualitative approach, studying outsourcing 'from the inside', 'in the making', which provides a novel perspective. Following recent developments in process studies [12], they offer situated and dynamic analyses based on vivid, rich and longitudinal data which are attentive to work activities, time and temporalities, and boundaries. They study the daily practices and interactions between the principal and the contractor(s) in the light of the lived experience and everyday working lives of those involved but also of longer-term institutional, professional and/or contractual arrangements. Many of the chapters also simultaneously study both parties and their inter-relations rather than favouring one actor's point of view over another's. This approach enables, as the findings put forward in the chapters show, development of a nuanced and balanced approach of outsourcing and of its implications for safety. This leads to the conclusion that outsourcing is neither always bad nor good, but that the effect on safe industrial performance depends on the outsourcing situation, the quality of interactions between the principal and the contractors and how these interactions are conceptualised and supported or not by organisational or socio-material devices. A central debate regarding this is the balance organisations should seek between specialisation and standardisation on the one hand and adaptation and informal practices on the other hand.

Also, the different chapters present in the book address a wide range of activities, some that are often studied (e.g. maintenance or design) and some that are less represented in the research literature (production, i.e. core business activities, or regulation/inspection). This reflects an evolution regarding outsourcing, which is no longer limited to peripheral activities, but which concerns activities central to system safety and also how safety is controlled and regulated.

1.3   The Structure of the Book

This book invites readers to plunge into the diversity of outsourcing practices and explore how they emerge, develop, change or stabilise over time, with a focus on how safety is affected by these practices and conversely. As Hopkins encourages us to do, we address the way operating companies, contractors and subcontractors manage 'to do things' safely in different settings and organisational forms, from permanent to temporary structures and from operations or maintenance to governance activities. Thanks to this variety, this book paves the way to a comprehensive discussion of these complex and important issues by the combination of traditional and emerging views.

The first three chapters deal with outsourcing in the context of ongoing operations, two in the nuclear sector and one related to railways. In these cases, contract workers need to work alongside workers employed by the principal.

Tillement and Leuridan study an outsourcing process in the making at a nuclear plant. They find that the organisational distance between those who make outsourcing decisions and those who are currently doing the work which is to be outsourced is such that the tendered scope is a poor reflection of the work that needs to be done by the successful contractor. Building tacit knowledge accumulated over time into a contract scope is very difficult but necessary to ensure safe operations into the future. Finally, they show that moving to an outsourced arrangement is a dynamic process which must be designed to 'enable the transmission of skills and the renegotiation of practices and professional roles, which are necessarily built collectively and over a long period of time'.

Dechy and Largier warn that despite clear issues, 'the safety debate cannot be reduced to be for or against subcontracting'. They look at workload planning for outages at French nuclear power plants and the impact on maintenance contractors, finding that there is a large gap between work as imagined (all work defined four months ahead of the shutdown with scopes fixed) and work as done (lots of new items added and resource levels significantly uncontrolled with in some cases twice or three times the effort expended as originally planned). Workload planning is done with the aim of saving money, but in the end creates many short-term issues with work sometimes being done by people without the necessary qualifications in an attempt to meet schedule limitations. Contracting provides the flexibility to remove resourcing constraints, but some constraints are safety-related and so should not be abandoned.

Hara's chapter describes how contracting is organised for design and manufacturing of vehicles of the Japanese bullet train. Despite use of multiple contractors, relationships between the operator and the contractors are established over years and all short-listed contractors are expected to contribute to the design specification against which they all prepare tenders. This collective approach is in significant contrast to arrangements described in other chapters. Contractual arrangements effectively force designers to share their ideas while requiring manufacturers to take overall responsibility and to share their expertise with competitors. Such collectiveness would appear to have the potential to address some of the issues raised by others about short timeframes and lack of learning, but it is not without problems. The collective approach where interfaces are not strictly managed by specifications and similar mechanisms appears to have contributed to the only serious bullet train accident.

The next four chapters study cases of outsourcing for temporary organising linked to supply chains for capital projects.

Gotcheva et al. have investigated the potential for use of relational contracting with nuclear industry personnel in Finland. The poor project performance of current projects, as exemplified by major project delays, suggests that there is room for improvement in contracting approaches currently used. The overall view of industry partners was that incentivising contractors and proponents working together would be an improvement to the current adversarial arrangements fostered by turn-key contracts. The study shows that attitudes are beginning to shift to more trust-based approaches.

The chapter by olde Scholtenhuis compares properties of mindful organising with typical qualities of contracted work using buried utilities as a case study. He shows how qualities of contracted work (specialisation, transience and price competition) conflict with organisational qualities for mindfulness, thus working against system reliability and ultimately system safety. Recommendations to improve overall reliability are to reduce interfaces between supply chain organisations through integration mechanisms, create contractual incentives that reduce transience and separate direct construction costs from mindfulness-enhancing costs.

Using the case study of Norwegian energy construction, Helledal and Pettersen use the themes of time, task, team and transition to consider implications of temporary organising for workplace safety. Although they are focusing on worker safety, rather than system safety per se, their findings have much in common with olde Scholtenhuis. The focus on task in temporary organising at the expense of broader organisational considerations 'does not encourage understanding the "bigger picture" that working with safety often needs'. Acknowledging that project organising has advantages, they note that safety models do not allow for temporary organising. In their view, the solution is not incremental adjustments to existing safety theories but rather a stronger focus on the link between temporality and safety.

In capital projects, commissioning marks the handover of a new facility from the contractors who have designed and built it to those responsible for ongoing operations. This is the focus of Russel and Tillement's contribution. Their finding that projects benefit from involvement of operations personnel is not new, but their focus on the lived experience of both project personnel and operators gives them a novel perspective. They show how boundary objects and boundary spanners can address this longstanding issue.

The final three chapters describe cases of outsourcing in regulatory settings.

Use of consultants by a regulator is the subject of the chapter by Hayes, Chester and King. The inherent tension between economic and technical regulation of gas utility companies is magnified in this Australian case study by use of consultants. The lack of in-house technical expertise on the part of the economic regulator leaves them taking advice from consultants who they have effectively tasked with making recommendations that companies spend less on projects, including those with system safety implications. Using consultants amplifies the cost saving orientation of the economic regulator, even when it comes to spending on safety-related items. This is done supposedly in the public interest to reduce energy prices to consumers but leaves them vulnerable to long-term degradation of network reliability.

Eydieux's chapter uses the concept of organisational hypocrisy to study regulatory approvals processes for nuclear facility decommissioning. In the first case study, the operator outsources key activities and then acts in their safety demonstrations as if the work is done in house. In the other case study, the regulator contracts out technical analysis of the operator's decommissioning plans. The contracted experts come to the conclusion that the way in which the operator intends to contract decommissioning work is problematic for safety. This is characterised as organisational hypocrisy in that actors say that outsourcing is problematic for safety, and yet in both cases, they do it anyway. The chapter appeals for a more pragmatic discussion of outsourcing situated in field practice, rather than preconceived notions that contracting is always either good or bad.

The chapter by Naderpajouh, Zhang and Hayes takes a different perspective on outsourcing and regulation by considering the results of privatisation of building inspection. Using a public accountability framework, and drawing on the Grenfell Tower disaster, their chapter highlights the structural problems inherent in using private sector people to fulfil a public sector role. The solutions must therefore also be structural.

Finally, along the way, readers will hear different stories, each bringing specific empirical and theoretical contributions, but all providing grounded answers regarding the complex and nuanced link between outsourcing and safety. We hope that this collection of chapters can serve as inspiration for future researchers.

References

1. K.G. Abraham, S.K. Taylor, Firms' use of outside contractors: theory and evidence. J. Law Econ. **14**(3), 394–424 (1999)
2. M. Bourrier, An interview with Karlene Roberts. Eur. Manag. J. **23**(1), 93–97 (2005)
3. N. Dechy, T. Bourdeaux, N. Ayrault, M.-A. Kordek, J.-C. Le Coze, First lessons of the Toulouse ammonium nitrate disaster, 21st September 2001, AZF plant, France. J. Hazard. Mater. **111**, 131–138 (2004)
4. Deloitte, *How much disruption? Deloitte Global Outsourcing Survey 2020* (2020)
5. S.J. Doig, R.C. Ritter, K. Speckhals, D. Woolson, Has outsourcing gone too far? McKinsey Q. **2001**(4), 24–37 (2001)
6. C. Grusenmeyer, Maintenance: organizational modes, activities and health and safety. Use of a French national survey and in-situ analyses. Accid. Anal. Prev. **73**, 187–199 (2014)
7. A. Hopkins, *Disastrous Decisions: The Human and Organisational Causes of the Gulf of Mexico Blowout* (CCH, Sydney, 2012)
8. A. Hopkins, *Organising for Safety: How Structure Creates Culture* (CCH, Sydney, 2019)
9. N. James, *Report of the Inquiry into the Victorian On-Demand Workforce* (2020)
10. R. Johnstone, M. Quinlan, C. Mayhew, Outsourcing risk? The regulation of OHS where contractors are employed. Comp. Labor Law Policy J. **22**(2 and 3), 351–393 (2001)
11. A.L. Kalleberg, Precarious work, insecure workers: employment relations in transition. Am. Sociol. Rev. **74**(1), 1–22 (2009)
12. A.N.N. Langley, C. Smallman, H. Tsoukas, A.H.V. d. Ven, Process studies of change in organization and management: unveiling temporality, activity, and flow. Acad. Manage. J. **56**(1), 1–13 (2013)
13. M. Quinlan, P. Bohle, Job quality: the impact of work organisation on health, in *Job Quality in Australia*, ed. by A. Knox, C. Warhurst (The Federation Press, Sydney, 2015)
14. M. Quinlan, I. Hampson, S. Gregson, Outsourcing and offshoring aircraft maintenance in the US: implications for safety. Saf. Sci. **57**, 283–292 (2013)
15. J. Reason, *Managing the Risks of Organizational Accidents* (Ashgate, Aldershot, 1997)
16. C.T. Valluru, A. Rae, S. Dekker, Behind subcontractor risk: a multiple case study analysis of mining and natural resources fatalities. Safety **4**(40), 1–18 (2020)

Chapter 2
Contracting and Safety: Lessons from Observing an Outsourcing Process "in the Making"

Stéphanie Tillement and Geoffrey Leuridan

Abstract Drawing from an inductive study at a nuclear plant, this chapter provides insights about an outsourcing situation that is barely studied in the literature, i.e. an outsourcing process (1) "in the making" of (2) a production activity. We highlight how practical, professional and contractual arrangements are elaborated throughout the progressive transfer of the activity. We discuss, from a situated perspective, the positive and negative effects of outsourcing in the short term and draw attention to key lessons for ensuring safe industrial performance in the long term.

Keywords Outsourcing process · Production activity · Nuclear · Practice-based approach

2.1 Introduction

In a practice-based approach, safety is defined as "*a collective knowledgeable doing*" [10]. Far from being a state established once and for all, it is built through a dynamic process and "*performed in, by and through safety practices*" [9] realised by workers with different occupations, skills and responsibilities. Work in contemporary organisations is more and more distributed, not only within organisational boundaries, but beyond them, as a result of increased outsourcing and a greater use of contracting out [22]. In this context, and following the practice-based approach, ensuring safety or maintaining safe practices is a collective endeavour in which actors, who belong not only to different communities of practices but to different organisations, are involved.

Looking at the literature on outsourcing and safety, we find that the vast majority of studies identify a negative link between safety (either occupational or process) and contracting. In the oil and gas sector, Hayes and McDermott [13] have shown the potential adverse safety outcomes of subcontracting, due to trade-offs on safe work practices, in order to increase some aspect of industrial performance (project deadlines and profit). Similar findings, which emphasise the tensions and vulnerabilities

S. Tillement (✉) · G. Leuridan
IMT Atlantique, LEMNA, Nantes, France
e-mail: stephanie.tillement@imt-atlantique.fr

© The Author(s) 2022   9
J. Hayes and S. Tillement (eds.), *Contracting and Safety*,
SpringerBriefs in Safety Management,
https://doi.org/10.1007/978-3-030-89792-5_2

arising from the use of contracting out, have been observed in the nuclear industry [23]. In aviation also, Bağan and Gerede [1] highlight the hazards associated with outsourcing due to cost and time pressures on contractors. Contracting has also been highlighted as a possible cause of incidents [20] and even accidents, e.g. Deepwater Horizon [12]; Challenger [24]. Among the activities that are contracted out, maintenance is one of the most common, which is often justified by the fact that it falls outside of the core competencies of firms. Paradoxically, it is also an activity known for being directly linked to safety.

Yet, most of the research undertaken has not examined the way in which work is actually performed when part or all of an organisation's activities are contracted out. In fact, existing research generally examines the organisational configurations in which activities are contracted out over a period of time, meaning that the activities therein have stabilised and routines put into place. Developing an understanding of the process in which activities are contracted out per se, i.e. the way in which outsourcing is established in the first instance, appears to be a blind spot in the literature.

The researchers of the present chapter are well placed to address this gap, having the opportunity to observe the contracting out of a production activity within a high-risk organisation. Drawing from an inductive study of one production workshop in a nuclear plant, the objectives of this research are to answer the following questions:

- How is the outsourcing of a production activity performed?
- How does outsourcing affect safe industrial performance?
- What pitfalls or good practices may be revealed through the observation of an outsourcing process in the making?

2.2   Outsourcing, Contracting Out and Safe Industrial Performance

The 1970s were characterised by an upward trend in outsourcing, which was consistent with the strategy of organisations to focus on their core businesses. With the decision to "make or buy", contracting out became the adjustment factor: adjusting the production capacity or adjusting the type of activity that should be done internally or externally. In the 1980s, this strategic movement became more prevalent with the development of labour flexibility (part-time jobs, short contracts, etc.) and the development of inter-organisational relationship models [16]. The increased use of contracting in almost every industry raises the question of the relationship between outsourcing and industrial performance.

The performance induced by a decision to contract out activities is correlated with an organisation's capacity to protect its strategic resources and competencies to sustain a competitive advantage [2, 26]. According to the resource-based view, the impact of outsourcing on organisational performance depends on how organisations are able to outsource efficiently. According to Barthélemy [3], organisations should avoid the "deadly sins" of outsourcing such as contracting out activities that should not be, selecting the wrong (sub)contractor, establishing a poor contract, overlooking personnel issues, losing control over the outsourced activity, overlooking hidden costs and failing to plan an exit strategy.

The relationship between outsourcing and industrial performance is thus complex, and contracting out is not necessarily synonymous with increased performance [14]. It may improve performance depending on whether there are contractual difficulties related to the appropriation of a proprietary technology, the cost measurement, and the interdependencies between outsourced and internal activities. In summary, there is no direct relationship between performance and outsourcing. This depends on firms' ability to develop competitive advantages by adapting to market conditions [6]. In high-hazard settings such as the nuclear industry, the ambiguous relationship between outsourcing and industrial performance cannot be considered separately from safety requirements.

With regards to Occupational Health and Safety (OHS), most studies highlight the negative impact of outsourcing on workers [19]. For example, an Australian study shows that contracted workers have a mortality rate over twice that of employee workers [17]. This rate can be explained by the fact that the most hazardous activities are often left to (sub)contractors. Moreover, outsourcing increases the complexity of work organisation—multiple stakeholders from different organisations, compartmentalisation of activities, co-activities, poor communication, etc.—and tends to lead to work intensification due to economic pressure and competition between subcontractors [17]. In addition, (sub)contractors are not necessarily qualified or well informed about risks [8]. In the French nuclear industry, contract workers absorb 80% of all radioactive doses received during maintenance activity [23]. The fragmentation of activity between the industrial operator and (sub)contractors hinders the measurement of work accidents [7].

Beyond OHS, establishing a causal link between the increasing use of subcontracting and a negative impact on safety is complex [20]. Maintenance in civil aviation and in the nuclear sector has been extensively studied. For example, Belobaba et al. [4] demonstrate the absence of statistical significance between contracting out and safety in civil aviation maintenance. However, outsourcing tends to generate "latent failures" [21]. Notably, increased cost pressures can lead to dangerous practices [1]: lack of training or skills, poor information about risks, co-activity, etc. By increasing work distribution, outsourcing exacerbates the problem of managing interdependencies and interfaces [11], which is crucial for safety since most risks lie at the interstices [5]. Both the principal organisation and (sub)contractors must manage these problems. In the French nuclear industry, the Environment Act prohibits delegating the responsibility for safety to a contractor. Industrial operators must monitor safety–critical activities when they are contracted out and ensure that contractors have the necessary technical skills to carry out these activities.

High-hazard organisations cannot favour performance at the expense of safety or vice versa. On the contrary, they have to simultaneously meet safety and industrial performance requirements, and outsourcing may impact both dimensions. Yet, as exhibited above, most studies focus on one dimension or the other. Our chapter goes some way to addressing this gap by considering the link between outsourcing a core activity (in this case, production) and "safe industrial performance". As stated by Journé and Tillement [15], "[t]his concept aims to put emphasis on the articulation between the safety goal and the other industrial goals by focusing on the processes, actors and objects that make this articulation possible under good conditions."

With regard to the above, we aim to understand this articulation by studying a contracting out process in the making, through a practice-based lens [10]. This involves paying particular attention to collective working practices developed by all stakeholders engaged in the outsourcing process whether this is the principal or the contractors and to their interactions and arrangements. In the next section, we highlight some important findings drawn from a case study in a nuclear plant.

2.3   The Trajectory of an Outsourcing Process in a Nuclear Plant

Our findings are drawn from an inductive study conducted at NucCo (a pseudonym) in close collaboration with industry practitioners, an arrangement which provides a unique opportunity to gain long-term access to the field. NucCo is a nuclear plant that specialises in radioactive waste processing. This very large, complex and tightly coupled [18] organisation faces the challenge of managing safety–critical activities. Yet, its safety objectives cannot be achieved at the expense of industrial performance (and vice versa). Approximately half of the personnel working for NucCo are contractors. Work is distributed between several workshops, which are all interdependent with regard to the waste treatment chain.

Fieldwork is conducted in one of these workshops, hereinafter referred to as W', dedicated to the collection and conditioning of waste from the other workshops. More specifically, we focus on the production activity, which consists of managing the conditioning and storage of waste packages depending on their radiological properties. NucCo decided in 2016 to outsource production of W', and this activity was officially transferred to the contracting firm at the beginning of 2019.

The main objective of our case study is to analyse an outsourcing process in the making. Consequently, neither the work situations nor the formal and informal relationships and arrangements are well established or have stabilised. Observing this process is an excellent opportunity to understand good practices and pitfalls when outsourcing an activity. More importantly, the authors negotiated from the beginning to have similar levels of access and proximity to the workers of the principal and contracting companies in order to analyse both sides of this outsourcing project.

Through a dynamic and situated approach, we collected data through observation (work in the control room, planning meetings, operational meetings) and interviews with both the employees of the principal and of the contractor. This study enables us to examine, throughout the outsourcing process, the evolution in the nature and quality of interactions between the workers of both companies, how work is performed, how contractual arrangements are negotiated and how this affects safe industrial performance.

2.3.1 The Issue of Defining Precisely the Outsourced Activity from the Beginning

Our study first reveals an incomplete characterisation of the activity to be outsourced from the very beginning. When top management decided to outsource this activity, it was based on the (partially flawed) assumption that this activity was simply about managing waste, similar to a nuclear clean-up. This definition led them to choose a contractor that had developed this area of expertise and was already working in other workshops of the plant. Yet, as the outsourcing process was unfolding, they realised that the activity was more about workflow management than clean-up, and thus about ongoing operations rather than one-off interventions. With the facilities being more and more monitored directly by the operators of the contractor, they found themselves confronted with volatile and unstable work, a situation far removed from the repetitive and routine job that was initially envisioned. They realised that the practices, associated knowledge and skills needed to manage the production activity safely and efficiently differed from those envisaged initially. This led to twofold tensions between the principal and the contractor: occupational and contractual. On a professional basis, the workers had to collectively engage in a process aimed at redefining the outsourced activity. This required rethinking the occupational roles of the operators and management staff of both the contractor and the principal, and to circumscribe their respective field of responsibilities. This had to be translated into contractual terms, leading to renegotiations of the initial contract to formalise the responsibilities and tasks of the external company to properly reflect the real activity.

More importantly, our study reveals how the workers had to progressively adapt their practices, both formal and informal, when confronted with the reality of the required activity. These practices have direct effects on safe industrial performance. This is detailed in the next section.

2.3.2 The Trickle-Down Effects of Outsourcing on Safe Industrial Performance

Our study suggests a nuanced vision of the link between outsourcing, safety and industrial performance.

Throughout the outsourcing process, the necessary practices and skills (largely informal and tacit) for carrying out production are gradually transferred from former internal operators and managers to the new external ones. This is supported by formal practices (writing or update of operating procedures) and more informal ones, notably inter-organisational forms of mentoring between experienced operators and newcomers. Paradoxically, this can be seen as a positive effect of outsourcing, as it precipitates a process whereby previous forms of normalised and often opaque or undiscussed "ways of doing" are revealed and questioned. The actors engage in collective discussions regarding these practices, with two possible outcomes. Firstly, the previous or inherited practices may be labelled as forbidden in the context of contracting out. Or, the practices may be authorised: in that case, they are formalised in operating procedures. In both cases, the contract is modified to formalise the agreement on the new scope of responsibilities of the contract workers.

Yet, as it is still not possible to anticipate and prescribe solutions for every possible event, vigilance in the management of "grey" areas should still be applied. Conducting work safely and efficiently requires the organisation to engage in an ongoing search to achieve a balance between control and autonomy. These tensions translate into two different, yet required, forms of commitment: contractual and professional. The second is linked to tacit knowledge and experience as acquired through confrontation with the materiality of installations. Acquiring autonomy is thus a long-term endeavour for contractors.

2.4 Conclusion: Lessons Learned

Our practice-based approach of studying an outsourcing process in the making enables a nuanced view of the link between outsourcing and safe industrial performance. It offers insights into the organisational and professional conditions that affect the positive or negative impact of outsourcing on safety and industrial performance. Finally, several lessons can be drawn from this case study with regard to enhancing safe industrial performance when outsourcing core and risky activities.

2.4.1 Lesson 1: A Correct Assessment of the Nature of an Activity Before Outsourcing Is Essential

As obvious as it may seem, it is of major importance to precisely assess the nature of the activity that a firm wants to outsource. In practice, this is not an easy task. As highlighted by Perrow [18] in his analysis of error-inducing systems, the "elites", in our case those who actually make the decision to contract out, are often far removed from the production or operating system and thus have incomplete knowledge of the actual work carried out by the actors in the field. Precisely assessing the nature of the activity (and the concrete practices and tacit knowledge that its execution requires) demands a full appreciation of relevant operations [25]. In a situation where the system is not fully in place, engaging with and seeking feedback from those who monitor, on a daily basis, the current state of the system and the related operations and practices that occur, is essential.

2.4.2   Lesson 2: Hidden Temporalities and Embracing Distant Past and Future Are Key in the Contracting-Out Process

Our study highlights the importance, but also the difficulty, of taking into account the hidden temporalities during an outsourcing process. These hidden temporalities are directly correlated with the nature of the outsourced activity. They refer to the distant past (informal practices that have been normalised and have become largely invisible and tacit over time), to the uncertain future of the acquisition of skills and knowledge and to the discrete timeframe of the present activity, which, in large part, consist of managing for the unexpected which cannot be fully planned for. These hidden temporalities, although essential to the efficient and safe performance of the activity, are difficult to integrate into the contractual logic that dominates any outsourcing process.

2.4.3   Lesson 3: Contracting Out Is a Dynamic Process, Which Requires Heedful Interactions Between Principal and Contractor in the Long Term

An outsourcing process is fundamentally dynamic. Far from being a simple task of moving directly from state A to state B, it requires processes that enable the transmission of skills and the renegotiation of practices and professional roles, which are necessarily built collectively and over a long period of time. The quality of interactions between management and the operators, and between the employees of the principal and the contractor, is key to maintaining vigilance throughout the process and reaching an agreement on good practices and the jurisdiction of each actor.

The situation that we observed (i.e. when production was being outsourced) does not allow us to deduce what the future situation will be. Ensuring safe industrial performance in an outsourcing context is an enduring endeavour, and vigilance should be maintained in the long term. At the time when the activity is outsourced, the transfer of activity is vigilantly monitored and supported by mentoring practices and formal meetings and evaluation. They reveal and enable us to collectively discuss the validity of practices that may have been out of step and largely opaque until then. Yet, as Vaughan [24] has shown, it is maybe when things tend to stabilise that a process of normalisation of deviance can gradually take place. Heedful interactions and discussions between the principal and the contractor should be maintained in the long term, else there will be a risk that forms of organisational blindness occur as a result of the long-term effects of organisational transformations.

Acknowledgements This work was supported by the RESOH Chair. We are very grateful to field workers for their time and information.

Ethical Statement This work adhered to the research ethics that are stipulated in the "RESOH Chair convention" that complies with relevant legislation regarding ethical conduct of research. Informed consent was obtained from participants, and all data have been anonymized.

References

1. H. Bağan, E. Gerede, Use of a nominal group technique in the exploration of safety hazards arising from the outsourcing of aircraft maintenance. Saf. Sci. **118**, 795–804 (2019)
2. J. Barney, Firm resources and sustained competitive advantage. J. Manag. **17**(1), 99–120 (1991)
3. J. Barthélemy, The seven deadly sins of outsourcing. Acad. Manag. Perspect. **17**(2), 87–98 (2003)
4. P. Belobaba, A. Odoni, C. Barnhart, *The Global Airline Industries* (Antony Rowe Ltd, Chippenham, 2009)
5. M. Bourrier, An interview with Karlene Roberts. Eur. Manag. J. **23**(1), 93–97 (2005)
6. O. Bustinza, D. Arias-Aranda, L. Gutierrez-Gutierrez, Outsourcing, competitive capabilities and performance: an empirical study in service firms. Int. J. Prod. Econ. **126**(2), 276–288 (2010)
7. V. Daubas-Letourneux, A. Thébaud-Mony, Les angles morts de la connaissance. Travail et Emploi **88**, 25 (2001)
8. T. Dwyer, *Life and Death at Work: Industrial Accidents as a Case of Socially Produced Error* (Springer Science & Business Media, New York, 2013)
9. S. Gherardi, *Organizational Knowledge: The Texture of Workplace Learning* (John Wiley & Sons, Oxford, 2006)
10. S. Gherardi, A practice-based approach to safety as an emergent competence, in *Beyond Safety Training* (Springer, Cham, 2018), pp. 11–21
11. C. Grusenmeyer, *Les Activités de Maintenance. Exploitation d'une Enquête et Analyse Ergonomique dans une Entreprise*, Note technique, INRS (2013)
12. J. Hayes, A. Hopkins, Deepwater Horizon—lessons for the pipeline industry. J. Pipeline Eng. **11**(3), 145–153 (2012)
13. J. Hayes, V. McDermott, Working in the crowded underground: one call services as a boundary object. Saf. Sci. **110**, 69–79 (2018)
14. R. Jarmon, A.S. Paulson, D. Rebne, Contractor performance: How good are contingent workers at the professional level? IEEE Trans. Eng. Manage. **45**(1), 11–19 (1998)
15. B. Journé, S. Tillement, La Gestion de la Sûreté dans le Nucléaire, in *Organisation, Information et Performance* (Presses universitaires de Rennes, 2016), pp. 175–185
16. A.L. Kalleberg, Nonstandard employment relations: part-time, temporary and contract work. Ann. Rev. Sociol. **26**(1), 341–365 (2000)
17. C. Mayhew, M. Quintan, R. Ferris, The effects of subcontracting/outsourcing on occupational health and safety: survey evidence from four Australian industries. Saf. Sci. **25**(1–3), 163–178 (1997)
18. C. Perrow, *Normal Accidents: Living with High Risk Systems* (Basic Books, New York, 1984)
19. M. Quinlan, P. Bohle, Under pressure, out of control or home alone? Reviewing research and policy debates on the OHS effects of outsourcing and home-based work. Int. J. Health Serv. **38**(3), 489–525 (2008)

20. M. Quinlan, I. Hampson, S. Gregson, Outsourcing and offshoring aircraft maintenance in the US: implications for safety. Saf. Sci. **57**, 283–292 (2013)
21. J.T. Reason, *Managing the Risks of Organizational Accidents* (Ashgate Publishing Ltd., Brookfield, 1997)
22. D. Tazi, *Externalisation de la Maintenance et Sécurité: Une Analyse Bibliographique. Les cahiers de la sécurité industrielle, Institut Pour une Culture de Sécurité Industrielle* (2010)
23. A. Thébaud-Mony, C. Levenstein, R. Forrant, J. Wooding, *Nuclear Servitude: Subcontracting and Health in the French Civil Nuclear Industry* (Routledge, New York, 2017)
24. D. Vaughan, *The Challenger Launch Decision: Risky Culture, Technology, and Deviance at NASA* (The University of Chicago Press, Chicago, 1996)
25. K.E. Weick, K.M. Sutcliffe, *Managing the Unexpected: Resilient Performance in an Age of Uncertainty* (John Wiley & Sons, San Fransisco, 2011)
26. B. Wernerfelt, A resource-based view of the firm. Strateg. Manag. J. **5**(2), 171–180 (1984)

Chapter 3
Workload Planning Management of Maintenance Activities in Nuclear Power Plants: Compensation Mechanisms at the Contractor Interface

Nicolas Dechy and Alexandre Largier

Abstract By reviewing data from a normal operations safety assessment conducted by IRSN on nuclear safety management, we study some factors that impact the management of maintenance activities. We focus on practices, their benefits and limits, especially in the case of drifts that lead to compensation mechanisms observed at the contractor interface when dealing with workload planning.

Keywords Workload · Planning · Maintenance · Contractor · Safety

3.1 Introduction

Several accident investigations and research studies have highlighted the possible adverse effects of contracting. The general trend of "core business" refocusing and frequent organisational changes raises recurring concerns about management and governance of such risks. The contracting issue often becomes societal and controversial, with trade unions and NGOs highlighting transfers of responsibility for risky activities. However, the safety debate cannot be reduced to being for or against contracting. Moreover, the many forms contracting can take invite discussion about contracting in the plural [17]. In order to enhance organisational safety, the challenge is to study the conditions, practices and influencing factors that are required to obtain safe performance from industrial processes and prevent accidents.

Our goal is to describe and analyse some practices, either positive or negative, to organise maintenance work at the contracting interface. They are inferred from a normal operations case study, looking at day-to-day practices that may or may not be safe enough. Evidence is collected from a safety assessment that Institut de Radioprotection et de Sûreté Nucléaire (IRSN) performed on French nuclear power plant (NPP) maintenance outages and contracting management; these kinds of assessments are dedicated to providing evidence for safety regulation [13].

N. Dechy (✉) · A. Largier
Institut de Radioprotection et de Sûreté Nucléaire (IRSN), Fontenay-aux-Roses, France
e-mail: nicolas.dechy@irsn.fr

© The Author(s) 2022
J. Hayes and S. Tillement (eds.), *Contracting and Safety*,
SpringerBriefs in Safety Management,
https://doi.org/10.1007/978-3-030-89792-5_3

The analysis will look at the provisions made by the nuclear operator and contractors for workload planning, the practices, their benefits and limits. It will highlight the limits of anticipation for workload planning, the adaptation mechanisms to compensate for drifts and their side effects. The discussion will address the effects of the contractor interface on anticipation and adaptation mechanisms.

3.2 Maintenance Workload Planning at the Contractor Interface

In the safety literature, there are several ways to study and produce knowledge on organisational reliability, resilience and safety in high-risk industries. Some researchers advocate the study of normal operations [1, 12], rather than the study of accidents, failures and vulnerability [16]. They have been seeking factors, practices and "best ways" of particular relevance for explaining how success is obtained in adverse conditions, highlighting features of organisational reliability, resilience and safety. These performance levels emerge from, and are embedded in, the day-to-day work practices of a long chain of actors inside and outside of an organisation, which all have a history. The many goals that an organisation pursues lead to tensions and dilemmas, which require compromises that are always contingent. Also, when investigating practices, it seems reasonable to take an approach looking at positive and negative effects [14]. This is even more relevant for maintenance, with its dual effects on safety.

3.2.1 Anticipation and Adaptation

Many influencing factors impact maintenance performance [10]; here, we focus only on one of them—the workload planning activity. The workload planning activity involves finding a balance between requests for maintenance tasks within a time window and deadline, and the availability of human resources. In the context of a scheduled outage for maintenance of an NPP (see Sect. 3.3), the workload related to maintenance activities dramatically increases, which makes this process even more critical to outage performance. The time dimension remains key for planning maintenance work, and planning is related to production pressures which impacts available human resources and maintenance working conditions. Workload planning is a process which closely interacts with programming, preparing, scheduling and resource management. To us, it is less tangible than scheduling [20] and seems to have received less attention in the safety literature.

Though planning is a key feature of management, reliability, safety and resilience, its ordinariness may lead to a lack of attention or interest [20]. In fact, this key coordination process is rather complex, integrating several dimensions, actors, and purposes [19]: prescribing, anticipating, coordinating, managing, warning, exploring, verifying, reporting, legitimising, justifying and making official information. It is dynamic with regards to information flows and performance feedback. It can be "tightly coupled" [16] and lead to a roll-on effect [3] when a small delay can lead to a bigger slip, as observed on the NASA space shuttle flight and maintenance management program.

Moreover, in contrast to standardised chain production, maintenance outages face high rates of unexpected events, variability and uncertainty [6]. When uncertainty is high, planning can drift [11] to a "fantasy plan" [2].

High-reliability organising calls for the management of tensions and contradictions between high levels of planning and rule-making, and adaptation to the unexpected through improvisation [21]. The anticipation–adaptation/resilience combination has been regularly addressed [1, 12, 21]. A key contributor to adaptation and resilience is "organisational slack" [5], summarised as an excess of resources.

However, a practical and theoretical issue arises from the recurring gap between planned and observed workload, at the end of the outage in our case. Actors have already experienced failures of foresight and delays in previous outages, so what should they do to better anticipate, or adapt and compensate when they occur? As this is not only a performance problem but also a safety concern, how is safety considered in workload planning?

3.2.2   Effects of Contracting

Maintenance workload planning is highly dependent on the reliability of the information exchanged at the contractor interface. The question we add here is transverse to the previous questions: how much does the contracting interface impact processes of anticipation and adaptation for workload planning?

Coordination and cooperation processes involve interfaces and boundaries between actors and stakeholders creating a "space in between" e.g. [18]. Some accidents highlighted [7, 8]: communication barriers (e.g. Challenger launch), less frequent and shorter interactions, formalisation of relationships, cultural gaps, transfer of responsibilities, and lack of sharing of lessons learned. All these factors could lead to a more fragmented and virtual organisation, with more diverging interests. Contracting companies challenged that contracting was the weak point of the supply chain only adding risks, highlighting that they also added benefits [4].

Workload planning is the responsibility of the contractor, who is required to perform a service with a result. Under European law, the nuclear operator cannot explicitly require a number of workers to be available when it pays a fixed price. Accordingly, the contractor does not have to disclose its resource plans. A research paper [9] on NPP maintenance outages highlighted how much the involvement of the contractor was different from the customer's when negotiating deadlines at the business level. In addition, when schedule delays occur, four different kinds of decisions lead to a consumption of the workers' time: the decision to use their workforce, to put them on standby for a waiting period, to ask them to work overtime and to modify their working week [9].

3.3   Research Context, Method and Data

Maintenance of NPPs is a complex activity. There are thousands of equipment items. Scheduled outages for maintenance activities occur approximately annually and typically extend over several weeks, allowing 3–15,000 activities to be completed and mobilising hundreds of nuclear operator workers, as well as hundreds of contractors in dozens of contracting and subcontracting companies.

To manage these maintenance activities, two key management processes are implemented by the nuclear operator: outage and contracting management. Outage management is ensured by a project management team of around 50 nuclear operator actors, mainly coming from maintenance and operation departments, dedicated for six months to a year. These actors are in charge of the preparation, execution and coordination of the activities planned for the outage. In France, around 400 companies involving more than 22,000 workers help to maintain the fleet of 58 reactors, with around 50 outages per year. Ten thousand workers out of the 23,000 employed by the French nuclear operator manage maintenance activities. The contractors must be certified as competent and financially sound by the nuclear operator to compete for contracts. Maintenance of NPPs is a regulated activity.

To describe the set of maintenance workload planning-related practices implemented by both the nuclear operator and contractors, especially those related to anticipation, adaptation and compensation when drifts and unexpected events occur, we used the data reported to the French nuclear safety authority (ASN) in the IRSN report [13] on assessment of management of safety of activities contracted by the nuclear licensee. This assessment relied on 50 days of field observations, 150 interviews on three 2013 and 2014 outages in three NPPs and the review of hundreds of documents. It required several person-years of effort over two to three years by eight human and organisational factors (HOF) specialists and their supervisors, in which we were involved as specialists.

3.4   Findings and Analysis

3.4.1   Preparing for Maintenance Outages

Workload planning depends on previous outage preparation activities, but also on feedback from later activities, and on the contracting process. Specifically, the list of activities to be performed, called the maintenance program, can be defined up to ten years in advance and is updated regularly and "frozen" six months before the outage to facilitate the preparation of thousands of activities on a stable basis. It also aims at giving contractors some visibility of the workload and type of activities that will be contracted. During the last months of preparation, it is updated every week and during the outage every day, in both cases to integrate new demands (e.g. unexpected equipment deficiency).

The principle of the outage process is "to integrate contractors as early as possible during preparation". The nuclear operator has developed contracts that last five years (extending to six or seven) to enable contractors to invest in and secure their technical and human resources in the long term, based on a promise of a minimum level of workload. A four-month milestone before the outage for signing contracts and work orders has been established as a rule for years. On this basis, resource planning can start with the skills required and the number of working hours needed per activity. Workload planning then depends on the more detailed schedules to be established later in the last months of outage preparation, when the working windows are established.

As of 2013, the French nuclear operator had experience of more than 1500 outages for its 58 reactors since the 1980s. So, most management rules and good practices are known and formalised and should "secure" anticipation of workload and visibility for contractors.

3.4.2 Implementation of Provisions Related to Workload Planning

Evidence shows that the nuclear operator does not always comply with its own milestones and rules. The freezing of the program six months before the outage is not respected. At one site, a technician in charge of valve maintenance described that ten days before the outage, he already knew that activity levels had risen and the workforce would be insufficient, but he did not communicate this to the contractor, instead proceeding as if the plan was still viable. On another site, it was reported that the practice is different, and contractors are advised when planned work changes so that they can adjust their workforce accordingly. When contractors are advised of changes before the four-month milestone, this may not be formalised. A valve maintenance contractor manager reported that he was notified informally of a volume upgrade but that work orders to formalise the change were not signed.

IRSN [13] noticed some progress compared to 2012 outages: the added maintenance activities dropped from 50% to between 15 and 20% in the six-month period before the outage, and from 50 to 30% during the outage. In the valve maintenance market, there is a chronic underestimation of volumes; at the end of the year, 25% to 40% more is done. The first reason is that the estimated volume of working hours is based on the most efficient contractor. Also, the 58-reactor nuclear fleet does not contract for all the volume but aims to keep some margins to allocate more work to the best contractors. In addition, as explained by a business manager in the valve maintenance market, the nuclear licensee corrects initial underestimated plans when (four-month milestone) they sign orders but it still underestimates the unplanned maintenance work. In the end, the final number of hours is as much as three times the estimate at the four-month mark.

With such enormous variations, late provision of final scope details hampers workload planning. A contractor manager claims at a meeting to have received only 8% of the documents of the programmed activities four months before the outage because of delays from the 2013 outages that impact the 2014 outages. Another contractor manager reports that documents for only 50% of maintenance activities were received at the beginning of the outage, which compromised workload spreading. The result is that contractors lose their trust in the schedule due to unplanned increases in activity, constantly changing priorities and the sense that everything is urgent.

3.4.3   Restoring the Workload–Resource Balance

These difficulties are known and do receive organisational response. Several compensation practices were observed to restore anticipation in a reactive manner, with some using the levers of the contractual relationships. In the first practice, some late work orders are issued with contract amendment and additional contracts until the maintenance work starts. Requests may be vague, such as having more staffing than the work order for instance on some weekends. If they are formalised, work orders may lead to paying for the service even if there is no activity. If requests remain informal, the contractor may add staff but faces uncertainty.

The second practice involves the overbooking of resources: this would become common practice, though varying from NPP to NPP, and would allow the outage project managers to build up resource margins or slack to cope with the uncertainty of volumes and dates and the lack of visibility by minimising their consequences. The third practice is redeployment between outages and NPPs: some human resources planned for an outage are transferred to another outage at the same NPP or another NPP, which requires negotiations and decisions at a higher level between the nuclear operator fleet managers and the NPP directors. A weekly meeting has been set up for that some trade-offs could lead to an outage being "sacrificed". The fourth practice is the reallocation of resources between work orders: resources earmarked for other activities can be mobilised by contract amendment; however, when reallocated to another contract with a different scope, this implies insurance problems.

Last but not least, the nuclear operator established a system to pay for waiting hours of contractor workers and set up for each NPP a team in charge of recording waiting hours. The goal is to avoid false notification and to check that the contractor demonstrates it cannot redeploy its workers to other activities.

3.4.4 Contractor Adaptability

The implementation of these compensation mechanisms has various effects. Though anticipation capacities are restored with reactive measures, this tends to rely on the adaptability of contractors and their willingness to take financial risks by organising to make additional resources available when they have no guarantee of compensation.

The first effect relating to the practice of overbooking of resources: contractors also overbook resources in order to create margins that will allow them to be responsive. However, in order for these reserves to retain their margin function, both the contractors and the outage project must conceal them from each other (e.g. delays in preparing operating documents not communicated to keep personnel on site). This withholding of information can degrade the mutual trust of the partners and the relationship can become a "game of bluff". When the service provider itself uses subcontractors, it may also want to maintain margins. There may thus be a cascading amplification of planning difficulties at each level of subcontracting. Notice that after Fukushima in France, a maximum of three levels of subcontracting became the new norm to reduce risk transfers along the organisational chain in response to criticisms from NGOs and trade unions on contracting side-effects.

The second effect is on the workers' time and flexibility of geographic location: contractors may have to reorganise the work of their teams, for instance, in staggered working hours and in overtime, but also make temporary changes of location, at extremely short notice. This may concern up to 15–20% of activities, according to the nuclear operator. For a contractor manager, the worst case factor is the delays that lead to peak loads and staggered shifts. A contractor manager has difficulty motivating his staff because he has to ask them to make efforts (e.g. staying available over the weekend), which sometimes prove to be unrewarded because of schedule delays.

The third effect has a direct impact on safety with changes of team composition: contractor companies reassign staff to cope with peak workloads. These reassignments can sometimes be made without respecting the skills constraints of the people involved, defined during the preparation process and recorded in the site organisation charts. The fourth effect involving the use of additional resources from other subcontractors: temporary workers may also be mobilised, which may threaten the skills available.

3.5 Discussion and Conclusion

NPPs seek to optimise the availability of their contractors' resources to ensure that work is completed on time, to reduce lost production and to control costs. The evidence presented above indicates that it is largely the contractors who must regulate the dynamic balance between effort required and resources. Anticipation of workload can be thwarted by a deterioration in the conditions (e.g. outage delays of the previous year) under which these resources can be deployed. Several compensation practices are implemented on both sides in response to degraded conditions in order to re-establish a form of regulation. The compensation mechanisms link the provisions, the drifts, the compensation practices and the effects.

Firstly, these compensation mechanisms respond to the intrinsic limits of any planning in complex systems with high rates of unexpected events [6]. Thus, recognition of this reality would require a better characterisation of these occurrences (e.g. frequency, severity of unexpected events, etc.) and the development of adaptive capacity beyond "slack" [5]. The building of this resilience capability should rely on a critical analysis of the implementation of compensation mechanisms and practices that can have adverse impacts on safety, which are currently underestimated and therefore not adequately managed. The evidence provided here could foster the debate and analysis.

Secondly, the contractual relationships impact the anticipation and adaptation capabilities. This complicates and rigidifies this organisational interface, by adding financial and legal dimensions to this working relationship [7, 8]. Thus, it may lead to a focus on financial management of workload planning failures with compensation for waiting hours; but this narrow focus makes business partners lose sight of the side effects on individual and collective skills and working conditions which directly impact safety. Another influencing factor is the inherent limits of transparency efforts that are being made on both sides to remedy the overbooking of resources. It is in each party's interest to retain room for manoeuvre, whether for anticipation or adaptation. These limits on transparency can affect the relationship of trust at each level of subcontracting and lead to a "game of bluff" that increases the risk of "fantasy planning" [2]. Thus, this contractual dimension increases the distance in the interstitial space of interfaces and makes this interstitial space more sensitive [18]. In these deteriorated working conditions, the workload planning and spreading become rather fuzzy.

Thirdly, these compensation practices, which call for adaptive capacities from contractors, are a source of tension and risks in some cases, which are increased by the asymmetric contractual relationship. Indeed, the contractual requirement of strong mobilisation of the contractors' employees can have a negative impact on the skills mobilised during field operations: fatigue and tensions induced by changes in working hours or location can reduce the vigilance of the workers; continuous changes in the schedule, with alternating high peak loads and waiting hours, generate demotivation and irritation, which can reduce the serenity and vigilance of workers; team recomposition can degrade collective and individual competence [15]; late provision of information to workers reduces their ability to prepare their work. The workers' individual time capital is the one that is negotiated and split by contractual stakeholders [9], and it is sometimes detrimental to the workers' well-being and safety. In such a case, the balance is not in favour of the contractor, which can lose the contract, or of the employee, who can lose his job.

Last of all, the asymmetrical contractual relationship is seen as a power asset for the customer. Failures tend to be attributed to the contractor, overshadowing the customer contribution that must provide adequate conditions. This situation generates a climate of blame that is detrimental to safety, as it blocks cooperation and learning. Both business partners should admit that successes and failures are their common responsibility.

Acknowledgements To our reviewers Sarah Fourgeaud, Franck Anner, Olivier Dubois (IRSN); Jan Hayes, Stéphanie Tillement.

Ethical Statement This work was approved by IRSN, the public institute that support the French nuclear safety authority. Data collection and reporting by IRSN experts is undertaken under a protocol that informs participants of the use of their testimonies and work observations after cross-examination, with anonymisation of interviewees and nuclear power plants.

References

1. M. Bourrier, The legacy of the high reliability organization project. J. Contingencies Crisis Manage. **19**(1), 9–13 (2011)
2. L. Clarke, *Mission Improbable: Using Fantasy Documents to Tame Disaster* (University of Chicago Press, Chicago, 1999)
3. C.A.I.B., *Columbia Accident Investigation Board Report*, vol. 1. (National Aeronautics and Space Administration and the Government Printing Office, Washington, DC, 2003)
4. Comité sur les Facteurs Sociaux, Organisationnels et Humains (COFSOH), *Pour une Contribution Positive de la Maintenance Sous-traitée à la Sûreté Nucléaire* (2017)
5. R. Cyert, J. March, Organizational factors in the theory of oligopoly. Q. J. Econ. **70**(1), 44–64 (1956)
6. N. Dechy, S. Thellier, J.-M. Rousseau, J. Pansier, F. Jeffroy, Management of unexpected situations during maintenance activities: controlling risks despite uncertainty, in *Proceedings of the λμ19 Conference* (Institut pour la Maîtrise des risques (IMdR), Dijon, France, 21–23 Oct 2014)
7. Y. Dien, N. Dechy, Process safety of virtual organisations, in *Proceedings of λμ18 Conference* (Institut pour la Maîtrise des risques (IMdR), Tours, France, 16–18 Oct 2012)
8. Y. Dien, N. Dechy, Les risques organisationnels des organisations fragmentées, in *Proceedings of Les Entretiens du Risque* (Institut pour la Maîtrise des risques (IMdR), Paris, France, 26–27 Nov 2013)
9. T. Globokar, Compromis temporels dans la gestion des projets. Le cas de la maintenance nucléaire. Revue Française de Gestion 2004/5 **152**, 81–96 (2004)
10. C. Grusenmeyer, Maintenance: organizational modes, activities and health and safety. Use of a French national survey and in-situ analyses. Accid. Anal. Prev. **73**, 187–199 (2014)
11. J. Hayes, A. Hopkins, *Nightmare Pipeline Failures: Fantasy Planning, Black Swans and Integrity Management* (CCH, Sydney, 2014)
12. E. Hollnagel, D.D. Woods, N.C. Leveson (eds.), *Resilience Engineering: Concepts and Precepts* (Ashgate, Aldershot, 2006)
13. IRSN, *La Maîtrise des Activités Sous-traitées par EDF dans les Réacteurs à eau Pressurisée en Exploitation, Rapport pour la Réunion du GPR du 11 Février* (2015)
14. B. Journé, Introduction, Chaire RESOH, Journée partenaires, Paris, France, 20 January 2015
15. A. Largier, C. Delgoulet, C. de la Garza, Quelle prise en compte des compétences collectives et distribuées dans la gestion des compétences professionnelles? PISTES 10–1 (2008)
16. C. Perrow, *Normal Accidents, Living with High Risk Technologies* (Princeton University Press, New Jersey, 1984)
17. M. Ponnet, Thèse de Doctorat, Les relations de sous-traitance et leurs effets sur la sûreté et la sécurité dans deux entreprises: SNCF et GrDF, Université de Nantes (2011)

18. K. Roberts, P. Madsen, V. Desai, The space between in space transportation: a relational analysis of the failure of STS-107, in *Organization at the Limit* (Blackwell Publishing, Malden, 2005)
19. S. Tillement, S. Gentil, Construction d'une performance industrielle sûre au sein de projets complexes, Chaire RESOH, Journée partenaires, Paris, France, 20 January 2015
20. S. Tillement, J. Hayes, Maintenance schedules as boundary objects for improved organizational reliability. Cogn. Technol. Work **21**, 497–515 (2019)
21. K.E. Weick, K.M. Sutcliffe, *Managing the Unexpected: Assuring High Performance in an Age of Complexity* (Wiley, San Francisco, 2007)

Chapter 4
Inter-organisational Collaboration for the Safety of Railway Vehicles: A Japanese Case

Takuji Hara

Abstract Shinkansen, the Japanese bullet train, has been operating for over 55 years without serious system accidents. Rolling stock is designed and manufactured through a collaboration between railway companies and contracted manufacturers. Safety is ensured through interactions among organisational actors, material entities, institutions and structures. Safety of such a large technical system in an inter-organisational setting cannot be built by material entities but also requires institutions for coordination, and the mindfulness and leadership of relevant actors.

Keywords Inter-organisational relationships · Safety management · Contracted manufacturing · Japanese railway · MAIS approach

4.1 A Serious Incident

At 1.33 pm on 11 December 2017, the No. 34 Nozomi Super Express (Shinkansen) left Hakata Station, bound for Tokyo. At approximately 1.50 pm, a conductor noticed an unusual smell, which was reported to staff at the control centre. However, due to problems in communication and decision-making, it took more than three hours and 740 km to stop the train and conduct an underfloor inspection, during which an oil leak was discovered. A detailed investigation revealed that one of the bogies had a large crack in its frame and was only 3 cm away from snapping completely. This means that the crack had put approximately 1000 passengers in danger for hours. The Japan Transport Safety Board (JTSB) declared this case as Shinkansen's first "serious incident".

On 28 February 2018, the train car operator, JR-West, explained that when Kawasaki Heavy Industries manufactured the bogie in 2007, they ground down the underside of one part of the frame to correct unevenness caused by imperfect processing. This then weakened that part and caused it to crack due to metal fatigue. The thinnest area of the bottom plate in question was 4.7 mm, in contrast to the

T. Hara (✉)
Kansai University, Suita, Japan
e-mail: t_hara@kansai-u.ac.jp

standard thickness of 7 mm that is listed in the specifications. According to a press release issued on the same day by Kawasaki Heavy Industries, company policy prohibited any bogie frame parts from being ground down thinner than the standard specifications, and a notice to this effect had been posted at the time. However, the worker responsible was unaware of the policy, and the foreman did not check the work after completion. The company's investigation also revealed microscopic flaws in the bogie frame, as well as the deposit welding on the primary spring seat. It is possible that the combination of these factors, together with the excessively thin bottom plate of the bogie frame, resulted in the crack [1, 2]. A JTSB progress report, released on 28 June 2018, also inferred that the weakening of the bogie frame part (due to the excessive grinding combined with the residual stress from the deposit welding) resulted in a fissure near the welds, triggering the development of a fatigue crack [3].

This case study explains that Shinkansen rolling stock is produced by contracted rolling stock manufacturers and parts suppliers, and operated and maintained by railway companies. This incident reveals a communication gap between the relevant organisations. In contrast, rolling stock design is conducted by multiple organisations but led by railway companies, as explained later. There are complex inter-organisational relations between railway companies and contracted manufacturers. This chapter will examine how railway vehicle safety is shaped within the context of these contract-based inter-organisational relationships and where problems with this process may lie, using an original analytic framework called "the MAIS approach".

4.2 The MAIS Approach

What is the MAIS approach? It originates from studies on "The Social Shaping of Technology" [4, 5].

"MAIS" is an acronym for "material entities" (M), "actors" (A) and "institutions/structures" (I/S). The MAIS approach is a research approach focusing not only on interactions between actors but also on interactions between actors and material entities or institutions/structures.

The purpose of the MAIS approach is to elucidate the processes of formation, reformation and transformation of human society under the constraints imposed by various material entities, institutions and structures, and to find clues for effective human intervention during these processes. The MAIS approach identifies actors involved in a social phenomenon. Furthermore, it elucidates his/her interactions with other actors or with various material entities and institutions/structures. Actors can be divided into individual and collective actors. They have unique interests, intentions, attributions of meaning, reflexivity and agency. They also attribute meaning to other actors, material entities and institutions/structures. Through their actions, they form, reform or transform other actors, material entities and institutions/structures. However, they are also constrained or reinforced by the actions of other actors, material entities and institutions/structures.

The MAIS approach also identifies material entities (both artificial and naturally occurring). Moreover, it elucidates the interactions among material entities, actors and institutions/structures. Different actors sometimes attribute different meanings to the same material entity. These entities are not only formed socially; they also possess physical properties (mass, energy, material stability, etc.), indicating that they form part of nature. Therefore, regarding material entities, it is also necessary to understand the physical properties involved in various interactions. Although actors attribute meaning to material entities and can control them to a certain degree, occasionally through the use of institutions/structures, they cannot have complete and permanent control over them because of their physical properties. Material entities not only constrain interactions among various actors, material entities and institutions/structures but also enable and facilitate them.

In addition, the MAIS approach identifies institutions/structures that strongly influence social phenomena. The approach clarifies how each institution and structure interacts with other institutions/structures, various actors and material entities. Institutions/structures are socially constructed throughout history. They are patterned social relationships that constrain or reinforce actors within a limited social space and for a certain period. Regarding the distinction between institutions and structures, the former are purposefully designed by certain actors and are often visible, whereas the latter emerge through the interaction of actors and are invisible. However, as this distinction is extremely complicated in practice, I refer to both collectively as institutions/structures. Institutions/structures constrain and promote interactions among other institutions/structures, various material entities and actors. However, they are also constrained or reinforced because of the said interactions. To understand the influence of these institutions/structures, it is necessary to understand the meanings attributed to each of them by various actors and the consequences (including unintended consequences) of their interactions.

Thus, the MAIS approach seeks to reconstruct the formation, reformation and transformation of social phenomena from the interaction of identified actors, material entities and institutions/structures, in order to thoroughly understand the social phenomena. In this case, we have applied the MAIS approach to the institutional system of railway vehicle construction in Japan drawing on data from two published papers [6, 7] and two interviews with individuals working for both railway companies and rolling stock manufacturers in July and August 2018.

4.3 Inter-organisational Relations and Efforts to Shape Railway Vehicle Safety in Japan

Japan's railway vehicles may be designed by railway companies or rolling stock manufacturers. Designs by rolling stock manufacturers can be further categorised into co-designs or one-company designs. Co-design is a process in which responsibility for the design is divided among multiple manufacturers. This is the method adopted by the Japan Railways Group (JR Group) including JR-West [7]. This chapter focuses on such co-design projects that involve inter-organisational interactions.

In a co-designed project, one or a few JR Group companies first create a system (basic) design of their new train vehicles. For Shinkansen vehicle development, there are cases in which only one JR Group company develops new vehicles as well as cases in which multiple JR Group companies are involved. They host the so-called *benkyokai* (study meetings) on the new train project with several rolling stock manufacturers, almost all of whom have a longstanding relationship with the railway companies. Such long-term inter-organisational relationships are common in Japan [8, 9]. Later, the JR companies assign the detailed design of different vehicles (power cars, passenger carriages, etc.) to each shortlisted manufacturer [6, 7]. The manufacturers are chosen by tender among the members of the *benkyokai*; usually, multiple companies are chosen because of the limited manufacturing capacity of each company and for risk management.

Each rolling stock manufacturer then holds another study meeting with their customers (JR company/ies) and several parts suppliers. Most of these parts suppliers have a longstanding relationship with the rolling stock manufacturer and the JR Group. Thereafter, a parts supplier is chosen as the designer of each component by tender. Such long-term and quasi-hierarchical relationships among suppliers are akin to the so-called *Keiretsu* [8, 9]. The order is placed by the rolling stock manufacturer or directly by JR company/ies. Parts suppliers prefer direct orders from the latter because this allows them to charge higher prices than when receiving an order from the rolling stock manufacturers. However, typically only certain important components, such as a major controller or a bogie, are directly ordered by the JR company/ies.

In this manner, the manufacturing and partial design of the train vehicles is outsourced to multiple rolling stock manufacturers and parts suppliers. Each contracted manufacturer is only responsible for its own part of the design, but when rolling stock is manufactured, the manufacturers must assume responsibility, not only for the elements they designed, but also for the elements designed by other companies. In other words, the outsourcing allotment of design and manufacturing differs [7].

In the railway vehicle development process of the JR group, the railway companies' vehicle design divisions decide on the vehicle concept, which includes its objectives and specifications. They also manage the overall project and reconcile the needs of manufacturers. They organise a *benkyokai* with potential contractors before orders, conduct design review meetings with contracted manufacturers for each allocated section and hold *tsunagi* (integration) meetings with almost all contracted manufacturers to secure the integration of the train's electrical and software systems. The head of such a division at a railway company is often the person in charge (PIC) of a vehicle's development [6, 7].

It should be noted that JR company/ies can collect many ideas about technology and safety from various potential contractors at a *benkyokai*. Manufacturers are so eager to accumulate a large order that they emphasise their technological excellence. Although manufacturers compete, their different ideas are incorporated into the design of a new train by the meetings [6].

In this manner, railway companies manage the outsourcing of the design and manufacturing of vehicles and parts. They seek to resolve issues, including safety issues, by exchanging information and opinions during preliminary consultations and at the basic planning stage, by holding design review meetings during the design process, and coordinating among external contractors. While long-term business relationships with multiple rolling stock manufacturers and parts suppliers do secure a socially and economically advantageous position for railway companies, these relationships also promote railway vehicle safety via the accumulation of railway-related expertise by the manufacturers and suppliers they outsource to [6]. This is the critical structure of inter-organisational relationships of co-design projects for railway vehicles in Japan.

During the manufacturing stage, collaboration between railway companies, contracted rolling stock manufacturers and parts suppliers continues. However, there is typically less collaboration between different rolling stock manufacturers at the production stage than at the design stage. As an exception, when a manufacturer faces a technical problem in manufacturing, it can consult its rival manufacturer or a parts supplier who has acquired experience or know-how on the matter. This type of relationship is reciprocal. However, at this stage, the relationships between railway companies, rolling stock manufacturers and parts suppliers are predominantly mediated by blueprints. There are few meetings to coordinate the different inter-organisational relationships of railway companies during the manufacturing stage. When manufacturing small and simple products, it is easy to standardise these through blueprints and automatic manufacturing equipment only. However, manufacturing processes for large products, such as railway vehicles and bogies, often require work to be done by hand. In these cases, social mechanisms and actions for the integration of relevant organisations and their workers are indispensable to ensure the safety of each product.

4.4 Analysis

We will now use the MAIS approach to analyse how vehicle safety is shaped in Japan's railway system. This is a complex sociotechnical system involving various inter-organisational relationships.

First, during the design stage, multiple actors (i.e. railway companies, rolling stock manufacturers and parts suppliers) pool their individual resources and knowledge and cooperate within the institution of co-design. This is achieved, even though there are potential conflicts among competing manufacturers, partly because of the structurally overwhelming power of railway companies as the sponsor and partly because train components exist only as symbolic information at this stage. At this stage, a railway company's vehicle design division (or the PIC within it) is an actor who takes the initiative to reconcile potential conflicts between manufacturers. The official reconciliation institution is the *benkyokai* and the design review meetings. The integration of inter-organisational relations during the design stage is thus achieved, not only by the efforts of the key actor (the JR vehicle design division), but also by institutional/structural factors, such as the power relationships between JR companies and the contractors and *benkyokai*. This is easier at the design stage than at the manufacturing stage because the products of design are symbolic information, which is copiable and free from material constraints of physical train components.

In terms of building safety into vehicle designs, *benkyokai* are used, wherein railway companies exchange information with rolling stock manufacturers and parts suppliers during the initial phase of the design to compile a list of anticipated problems and their solutions; the solutions are then incorporated into the design. The smooth functioning of this institution requires active commitments from all related actors: railway companies, rolling stock manufacturers and parts suppliers. Blueprints and various prototypes may be considered critical material entities, strongly related to incorporating safety into the vehicle design during the design stage. However, material entities only constitute a portion of the mechanisms for integration. They shape the safe design of railway vehicles together with the key actor's effort for integration and institutional settings such as *benkyokai*. Again, the backdrop of this unique collaboration among competing manufacturers is the JR-centric power distribution and JR's sense of responsibility and pride in the safety of the Shinkansen. The strong attitude towards safety of the most powerful actor and lower material constraints during the design stage than during the manufacturing stage seem to be the key for this inter-organisational collaboration for safety.

Collaboration between the actors of railway companies and contractors carries over into the manufacturing stage. However, there are far fewer interactions among relevant actors at the manufacturing stage than at the design stage. During the manufacturing stage, the relationships between rolling stock manufacturers and parts suppliers are mediated by material entities: thus, the rolling stock manufacturers assemble the parts manufactured by the parts suppliers according to the approved blueprints. Similarly, the relationships between rolling stock manufacturers and railway companies are also mediated by material entities, such as the rolling stock and blueprints. We also did not observe institutional devices for integration during the manufacturing stage, as with *benkyokai* in the design stage.

Another point of variance between the design and manufacturing stages is the increased influence of material entities that mass production entails. The institution of standardisation becomes critical when rolling stock, and parts are mass-produced. Standardisation of production processes is needed to ensure the standardisation of rolling stocks and parts. This requires not only the precision of material entities, such as blueprints and machine tools, but also the skill and focus (carefulness and/or mindfulness) of the relevant actors. The actions of workers are influenced by structural and institutional factors, such as the organisational culture of safety, rules and human resource management. It should be noted that this setting for the standardisation of railway vehicle production must be maintained consistently by all the relevant manufacturers of rolling stock and parts. For the safety of railway vehicles, the communication between relevant organisations should be stronger than dependence only on the material entities, such as blueprints. Institutions for the communication and commitment of relevant actors are also necessary.

In the serious incident described above, a portion of the bogie frame was weakened because the contracted rolling stock manufacturer used an inappropriate method to correct a defective part manufactured by a parts supplier. This happened because the worker responsible did not act in accordance with the standard policy of the production process. The defective part then cracked due to metal fatigue over time. As outlined above, one problem during the manufacturing process can affect the safety of the entire railway system. Shaping this safety requires smooth interaction among many actors (e.g. workers), institutions (e.g. organisational rules and cultures) and material entities (e.g. various structural parts). In this case, the sharing of information among relevant organisations (i.e. the manufacturer of parts, the manufacturer of the bogie and the railway operator who was responsible for the inspection and maintenance) was insufficient. During the manufacturing stage, there is no centralised actor analogous to the design stage's PIC. There is no institution designed to unify the actors analogous to the design stage's *benkyokai* and design review meetings. In addition, to produce a great number of heavy, complicated hardware items, such as railway vehicles and bogies, require a larger, dispersed and diversified inter-organisational setting, which makes coordination much harder than in the design phase.

4.5 Conclusion

Analysis using the MAIS approach revealed that there are actors (the vehicle design divisions of railway companies) and institutions (*benkyokai, tsunagi* meeting and design review meetings) in place for the overall safety coordination of the co-design of railway vehicles, but there are no such actors or institutions in place for their co-manufacturing. This is probably because the relevant organisations believe that making parts and vehicles exactly in accordance with the blueprints would achieve the correct results, without requiring further coordination. However, unlike design, manufacturing has to repeatedly mass-produce the exact same parts and vehicles; that is, manufacturing requires standardisation. This cannot be achieved by sharing the same blueprints. The manufacturing of railway vehicles requires handwork. Therefore, standardisation can be achieved not only by blueprints and correct manufacturing equipment but also by intentional and mindful efforts of the employees of contracted manufacturers and railway companies. Institutional devices for promoting actors' efforts are also crucial. When conducting inter-organisational manufacturing, we should create an inter-organisational institution for standardisation to secure product safety. We can increase the safety of railways by such inter-organisational institutions with shared organisational culture for safety, by the relevant actors' increased commitment to the cause and by mobilisation of material entities such as clear blueprints and correct manufacturing equipment during the manufacturing stage as well as inter-organisational coordination during the design stage.

We may be able to apply this to other contracting-based projects related to safety. When we produce a system in an inter-organisational setting, its safety cannot be built simply by material entities but also requires institutions for coordination, and the full commitment, communication and leadership of relevant actors. We should also create structural circumstances for safety and remove any structural factors against safety. We should find how to achieve these conditions practically even in the mass production of large, complicated and potentially dangerous systems with a dispersed and diversified inter-organisational setting. The MAIS approach can be helpful in analysing such a situation and identifying key material entities, actors and institutional and structural factors. However, further case studies of various projects are necessary for the validation of these conclusions.

Acknowledgements This work was supported by JSPS KAKENHI Grant Numbers 15K03567 and 20K01880.

Ethical Statement This work adhered to the research ethics that are stipulated in "For the Sound Development of Science-The Attitude of a Conscientious Scientist" edited by the Japan Society for the Promotion of Science (JSPS). Informed consent was obtained from participants and all data has been anonymised.

References

1. JR West, The Advisory Committee on the Serious Incident of Shinkansen, *Shinkansen ijo kanchi-ji no unten keizoku jisho heno saihatu boushi ni kansuru kento kekka ni tuite—shinkansen no saranaru anzensei koujo ni mukete [An Examination on the Measures Against the Recurrence of Keeping Shinkansen Running on the Detection of Abnormality—Towards the Improvement of Safety of Shinkansen]* (27 March 2018), https://www.westjr.co.jp/press/article/items/180327_00_yuushikishakaigi_2.pdf. Accessed 12 Aug 2018
2. Kawasaki Heavy Industries, *Notice of Series N700 Shinkansen Train Bogie Frames Matter* (News Release, 28 February 2018) http://global.kawasaki.com/news_C3180228-1.pdf. Accessed 12 Aug 2018
3. Japan Transport Safety Board, *Tetusdo jyudai inshidento chosa no keika houkoku ni tuite [Interim Report of the Investigation of the Railway Heavy Incident]* (28 June 2018), http://www.mlit.go.jp/jtsb/railway/rep-inci/keika180628.pdf. Accessed 26 Nov 2018
4. D. MacKenzie, J. Wajcman (eds.), *The Social Shaping of Technology*, 2nd edn. (Open University Press, Buckingham, 1999)
5. R. Williams, D. Edge, The social shaping of technology. Res. Policy **25**(6), 865–899 (1996)
6. T. Kitabayashi, Product development management of rolling stock in Japan and its global expansion: case study of T3 project. J. Japanese Oper. Manage. Strategy **6**(1), 34–54 (2016)
7. T. Kitabayashi, Collaborative factors among suppliers in multi-sourcing: case study of joint development for rolling stock in Japan. Nihon Keiei Gakkai-shi **39**, 3–14 (2017)
8. W.M. Fruin, *The Japanese Enterprise System* (Oxford University Press, Oxford, 1992)
9. T. Nishiguchi, *Strategic Industrial Sourcing: The Japanese Advantage* (Oxford University Press, New York and Oxford, 1994)

Chapter 5
Engaged Scholarship for Exploring Applicability of Relational Contracting to Nuclear Industry Projects

Nadezhda Gotcheva, Kirsi Aaltonen, Pertti Lahdenperä, and Soili Nysten-Haarala

Abstract We employed engaged scholarship as a research strategy for exploring the applicability of relational contracting in nuclear power projects. Insights from a series of workshops with nuclear industry practitioners in Finland indicated that although project alliancing is not a familiar contractual approach in the nuclear industry, the benefits of its implementation are increasingly recognised.

Keywords Relational contracting · Nuclear industry · Engaged scholarship · Finland · Project alliancing · Contracts

5.1 Introduction

Complex projects are temporal multi-organisational entities, in which the participating actors pool, integrate and coordinate resources, efforts, capabilities and knowledge to fulfil a unique common objective. Ensuring that safety and quality requirements are properly understood and satisfied in a multinational, oftentimes interdisciplinary and dynamic project context is a demanding and long-term process. Cost and schedule overruns are recognised as common performance problems in large complex projects. According to the World Nuclear Industry Status Report (2019), at least 59% of the 46 reactors currently under construction globally are delayed. In Finland, there are two new nuclear builds: Olkiluoto 3 nuclear power plant (NPP) was supposed to be operational in 2009, and at the moment of writing this chapter, it is still in pre-operational stage, while Hanhikivi 1 NPP was originally planned to

N. Gotcheva (✉) · P. Lahdenperä
VTT Technical Research Centre of Finland Ltd, Tampere, Finland
e-mail: Nadezhda.Gotcheva@vtt.fi

K. Aaltonen
University of Oulu, Oulu, Finland

S. Nysten-Haarala
University of Lapland, Rovaniemi, Finland

© The Author(s) 2022
J. Hayes and S. Tillement (eds.), *Contracting and Safety*,
SpringerBriefs in Safety Management,
https://doi.org/10.1007/978-3-030-89792-5_5

produce electricity in 2024 and is currently planning to get a construction licence in 2021.

The World Nuclear Association recently highlighted the need to enhance collaborative or partnership approaches by indicating that standard contractual arrangements may not be sufficient to ensure that interests of key stakeholders are aligned, and appropriate procurement and project delivery models are needed to support collaborative ways of working [25]. Relational contracting has been actively implemented as a collaborative approach for handling the complex challenges experienced in inter-organisational project networks, for instance, in the infrastructure construction domain. However, relational contracting is still not a familiar approach in the nuclear industry.

This study was motivated by the need to explore the applicability of relational contracting and inter-organisational integration in complex nuclear industry projects in Finland. Our assumption is that integration of some best collaborative practices from the project alliance type of contracting to turnkey contracting could be beneficial for nuclear industry organisations. The research question was: *What are the possibilities of applying relational contracting to improve the performance of complex nuclear industry projects?* We explored the challenges and potential benefits that could be captured from applying relational contracting in nuclear industry projects.

5.2   Contractual Approaches

Traditionally, a *contract* is understood as a tool to safeguard one's own interests against the interests of other contracting parties. This classical approach is drawn from contract law, the simplified model for a contract which reflects the simple sale of goods (purchase contracts), in which the interests of the seller and the buyer mostly indeed contradict each other [16]. Since contracts bind in their original form, this approach leads to drafting contracts which safeguard the drafter in a potential legal dispute [10, 11]. Such an inflexible "hard approach" in contracting may, however, lead to official contracts, which are locked in a safe box in case of litigation together with a more flexible business practice where the hard contract is circumvented or ignored when contingencies appear [21].

However, this is not the whole truth about contracting—not even in contract law. Freedom of contract is one of the main principles of contract law, which allows contracting parties to design their own agreement. Complex long-term contracts require a more sophisticated approach, for which a contract is not only a legal tool but also a tool for business cooperation [16].

5.2.1 Traditional Contractual Approaches in the Nuclear Industry

The procurement and delivery models adopted by the nuclear power industry vary. Traditionally, three main contractual approaches have been employed for the construction of new nuclear power plants [20]: (1) *Turnkey or Engineering, Procurement, Construction (EPC)*: a single contractor/consortium of contractors takes the overall responsibility for the construction work and delivering a complete and functional plant to the customer. The vendor or consortium may subcontract elements of the project, which it is not able to supply itself; (2) in *split package or hybrid,* the overall responsibility is divided between a small number of contractors, each coping with a section of the plant. In more complex split packages, the overall responsibility for design and licencing and for integrating the various packages should be allocated to either the plant's owner or one of the main contractors to ensure that plant's systems work jointly properly; (3) *Multi-contract*: either in-house or more often an external architect/engineering (A/E) company is responsible for the overall design, licencing, inviting bids and selecting contractors for plant's systems, managing the actual construction work as well as for testing and commissioning. The more there are separate components, the more challenging the A/E coordinator's task will be. In all variations of 2 and 3, it is important that there is either one or only a few main responsible partners representing the whole complex project.

5.2.2 Project Alliancing Approach

Project alliancing, sometimes also referred to as integrated project delivery, is a project delivery method based on relational contracting and a relationship of trust between multiple parties [13]. The project parties assume joint responsibility for the design and construction of the project to be implemented through a joint organisation, share both positive and negative project risks, and observe the principles of information accessibility and open book accounting [12]. The contract does not specify duties per party, but it determines the tasks needed to complete the project and all contracting parties assume a full responsibility for their fulfilment.

Project alliancing includes strong incentives for developing best-for-project mindset and culture, unanimous decision-making and commitment to no-disputes. "We all sink or swim together" is a common motto used when referring to the way of working in a project alliance. Usually, key service providers are involved early in relation to the design process, while their capability and collaboration ability are important selection criteria. A joint development phase for the development of project solution and fixing of the target cost precedes the implementation phase.

Project alliances have so far been mostly applied in the infrastructure and construction domains. Australia has been in a leading position in the development and introduction of the system [5, 23], which is used in other countries as well, with Finland being a forerunner as a country implementing "the pure" Australian approach [14].

Experiences in project alliancing have been mainly positive and belief in the excellence of the system is strong in general. However, an alliancing approach is not suitable for all projects. The Australian national guidelines [4] instruct that: first, this approach is suitable for high value projects due to the high initial start-up management costs; and second, it is a question of a challenging and risky project, when risks cannot be adequately defined prior to tendering, the cost of transferring risks is prohibitive in the prevailing market conditions, or a collective approach to assessing and managing risk will produce a better outcome. Some criticism of the "pure" Australian approach to project alliancing has also been presented [6].

5.2.3   A Legal Perspective to Relational Contracting

The term "relational contracting" is drawn from Stewart Macauley's famous article from the 1960s, in which he empirically proved how businessmen in Wisconsin made contracts based on trust. According to Macaulay, contractual relations are more important than legal contracts "signed and sealed" [14]. Alliance contracts are typical relational contracts because they require creating and maintaining trust between the parties [2, 23]. From a legal perspective [19], these contracts represent "soft contracting", which can be seen as the opposite to hard contracting (Fig. 5.1). Since most contracts include both hard and soft elements, this should not be seen as a dichotomy. For example, contracts in the nuclear industry typically include a lot of hard elements, such as mandatory safety regulations. However, strong safeguarding elements are often based on mutual cooperation.

Hard contracting	Soft (flexible) contracting
• Emphasises opposite interests of the parties • Highlights the need of the parties to safeguard themselves against risks • Uses precise and unchanging hard terms • Classic view of contract (legalistic)	• Emphasises flexibility, good will, commitment to cooperate with the parties • A framework for cooperation • Uses soft term agreement of reference or standards that can be specified later • Change mechanisms

Fig. 5.1 Hard and soft contracting (adapted with permission from [18])

The classic, legalistic view approaches contracts through disputes (cases), which have been decided in courts. However, over the last two decades, proactive law has suggested that academic contract law should also focus on contracts from an *ex ante* perspective, approaching contracts as tools for enabling business. Instead of focusing on past disputes, proactive lawyers prefer to prevent problems from arising or solve them creatively before they escalate into legal disputes [8, 9]. This is not done by safeguarding for every possible risk with precise, unchanging stipulations but creating and maintaining trust between contracting parties. This approach returns contracts to businesspeople, who are the real owners of contracts [15], and supports building a collaborative climate and relations and thus advances the use of relational contracting [1].

A contract can advise in communicating between different professionals participating in the implementation. It can function as a tool for coordinating and assigning roles and responsibilities of participants. It is a guidance in changing circumstances, and definitely, a contract should create value for the parties [17].

In the nuclear industry, safeguarding against potential accidents definitely is a major part of contracting. However, communicating to prevent things going wrong is certainly as important as stipulations on safeguarding rules. Locks and threats do not prevent risks without good communication to ensure that people know what they are expected to do. Soft elements from alliance contracts could turn a nuclear project into a joint project, in which benefits and costs are shared. If it is in everybody's interest to benefit and complete the project effectively on time, they all invest in cooperation and maintaining trust. Thus, sharing costs and benefits has a safeguarding function as well. The *no claim-no blame* clause, which lawyers often criticise, is a tool in creating and maintaining a team and an atmosphere of cooperation. With this clause, contracting parties agree not to take disputes before court or even arbitration. All disagreements should be agreed between the parties themselves with joint decisions. However, this clause cannot prevent litigation in the case of gross negligence. Therefore, soft elements replicated from alliance contracts alone can only strengthen nuclear projects by motivating all the key participants for reaching joint objectives in an atmosphere of cooperation and mutual trust. In the case of alliancing, the core is joint risk and liability, which aligns the interests of the contracting parties, and the no claim clause is a kind of a "cherry on top".

5.3 Method

As a methodological approach, *engaged scholarship* refers to collaborative engagement of academics and practitioners. This engagement is characterised as "a relationship that involves negotiation and collaboration between researchers and practitioners in a learning community; such a community jointly produces knowledge that can both advance the scientific enterprise and enlighten a community of practitioners" [23, p. 7]. Engaged scholarship points to the reciprocal relationship between academics and practitioners in terms of bridging the knowledge gap, and different types of reciprocity have been identified in project research contexts [7, 22, 24]. The "Scandinavian tradition of engaged scholarship" is characterised by investing a substantial amount of time in collaborating with industry partners and communicating results specifically to practitioners [22].

We designed and conducted a series of workshops in the Finnish nuclear domain to explore the possibilities of relational contracting in nuclear industry projects. We invited nuclear industry practitioners—representatives from the nuclear industry companies and the regulator in Finland—to the workshops. One of the workshops was international and cross-industrial, aiming at sharing insights from experiences with project alliancing and management of complex construction projects in Finland, Australia and the UK. The workshops collaboratively engaged researchers and practitioners from the industry and the regulator in discussing the possibilities of applying relational contracting to improve the performance in complex nuclear industry projects in Finland.

5.4   Results

Key insights from the workshops series in terms of potential benefits and challenges for applying relational contracting are presented as an illustration of engaged scholarship for exploring the applicability of this contractual approach to the nuclear industry. Regarding project alliancing and its application in the nuclear industry in Finland, it was highlighted by practitioners that in this industry it is very important for the actors to have independence, clear lines of responsibility and well-defined liability to ensure that the actors' roles do not become mixed. This relates to the fact that, in the nuclear industry, the licensee is responsible for safety. For example, the licensee needs to independently verify documentation prepared by other parties; yet, it was acknowledged that open communication, good relations and collaboration in the project are of utmost importance.

The practitioners considered it as very important to facilitate collaborative working on projects, and the safety regulator representatives particularly reflected on their role in promoting collaboration. One of the discussion points was how the current Finnish regulatory guides on nuclear safety promote or restrict a collaborative working mode in projects.

Co-locational collaborative project spaces that would enable inter-organisational collaborative working in the same physical space were considered as an interesting approach that would potentially facilitate knowledge sharing, collaborative practices and inter-organisational coordination. The safety regulator also saw this as a promising avenue to facilitate a relational orientation in everyday work but expressed concerns over confidentiality issues in co-locational spaces. The potential for the safety authority representatives to work in the same co-locational space was also discussed but challenges were seen to be related to this kind of arrangement due to the regulator's independent role. The practitioners, however, considered the benefits of the co-locational space to crystallise in the improved transparency and visibility of the work of others, as well as the increasing relational capital and trust among the project participants when they are working in the same premises.

The cultural diversity of the project participants was also a recurrent theme in the workshops. Although cultural diversity was considered to bring value to the project, cultural distance was seen to weaken the practical possibilities for relational contracting and collaboration. One reason for this was that different parties may have different understandings of what relational contracting and collaboration means in practice. This may also favour a strong impetus toward traditional contracting and safeguarding of issues as this is the practice that the parties are used to.

It was also brought up that the nuclear industry, as a highly regulated industry, has a strong controlling approach in terms of ensuring nuclear safety, preoccupation with failure and a tendency to view issues through the perspective of risks instead of opportunities. This, in turn, may favour the use of traditional contracting solutions and limit flexibility instead of using relational contracting approaches that emphasise more value co-creation and joint co-operation. The transformation towards the use of relational contracting is then also a significant institutional change effort which requires a lot of institutional work and questioning of the fundamental values shared by the industry actors. As practitioners noted, it is easier to talk about the need for collaboration in theory than to actually implement it in practice.

Practitioners also showed interest towards integrating some of the collaborative practices into traditional contracts and projects, introducing then some kind of hybrid contractual solutions that would support collaboration in practice. One potential area related to this was the introduction of bonus schemes and reward structures to the contracts that would motivate parties to share information, instead of a focus on penalty culture and sanctions in the contracts. Practitioners expressed concerns over the situation where the contracts do not encourage parties to share information in order to safeguard themselves and were considering solutions for improving, e.g. through statements on proactive information sharing and commitment toward that kind of culture.

The limitations of the current practice of turnkey contracting were recognised, too: for ensuring nuclear safety, in practice, the owner or licensee has to do more than formally required in the contract by supporting the contractor's duties. Especially when the suppliers are not very familiar with the local regulatory requirements, collaboration between actors needs to be intensified to ensure timely and proper understanding of the challenges and availability of information to support decision-making and delivery of high-quality safety documentation. If suppliers are not used to working according to rigorous regulatory requirements, this could make nuclear projects unattractive or even risky for them.

Regarding project alliancing and its joint liability, it was discussed that collaboration and risk sharing may encourage suppliers to tender since this delivery form creates favourable conditions for trust building and shared learning while mitigating the risks for suppliers. Flexible contractual approaches offer more room for adaptability in changing the project's goals/objectives. They set conditions for discussing and agreeing with suppliers, and offer novel ways to consider the suppliers' role in the project. All in all, it was recognised by the practitioners that relational contracting may bring benefits for overall performance of nuclear projects, and its application can be flexible too: for instance, relational contracting could be considered in subsections of a project with specific/strategic suppliers and not necessarily applied at the level of the overall project.

5.5 Conclusion

Large complex nuclear industry projects present significant managerial challenges as they seek to ensure a shared understanding of safety, develop common goals and achieve a good safety culture among the temporary network of participating organisations. Recent evidence from complex project research indicates that contractual approaches that promote the development of relational capital and trust-based social norms among project members are the most effective approaches to ensure the success of such projects [3].

In this chapter, we utilised engaged scholarship as a collaborative inquiry between academics and practitioners to explore applying relational contracting for improving project performance in the nuclear domain. EPC or turnkey contracting has been the traditional model used by the nuclear energy industry globally, which shapes a sort of path dependency and leaves little room for innovative contractual solutions. Yet, this study indicates that attitudes are changing and possibilities to improve the overall performance of nuclear projects via collaborative arrangements of relational contracting are increasingly being considered. Relational and proactive contracting approaches can be used side by side with more traditional approaches in the nuclear industry. Good relations between project actors are beneficial for preventing and mitigating dispute risks and misunderstandings and thus have implications for safety performance.

Acknowledgements The authors are grateful to participants of the series of workshops, organised as a part of project "Management principles and safety culture in complex projects" (MAPS), funded under SAFIR, the Finnish Research Program on Nuclear Power Plant Safety 2015–2018.

References

1. T.D. Barton, *Preventive Law and Problem Solving: Lawyering for the Future* (Vandepal Publishing, Lake Mary, 2009)
2. D. Campbell, D. Harris, Flexibility in long-term contractual relationships: the role of cooperation. Lean Constr. J. **2**, 5–29 (2005)
3. M. Chakkol, K. Selviardis, M. Finne, The governance of collaboration in complex projects. Int. J. Oper. Prod. Manag. **38**(4), 997–1019 (2018)
4. Department of Infrastructure and Regional Development, *National Alliance Contracting Guidelines: Guide to Alliance Contracting*, Australian Government, Department of Infrastructure and Regional Development, Canberra, Australian Capital Territory, 2015

5. Department of Treasury and Finance, *Project Alliancing: Practitioners' Guide*, Department of Treasury and Finance, Melbourne, Victoria, 2006
6. Department of Treasury and Finance, *In Pursuit of Additional Value: A Bench-marking Study into Alliancing in the Australian Public Sector*, Department of Treasury and Finance, Melbourne, Victoria, 2009
7. J. Geraldi, J. Söderlund, Project studies and engaged scholarship: directions towards contextualized and reflexive research on projects. Int. J. Manag. Proj. Bus. **9**(4), 767–797 (2016)
8. H. Haapio, Business success and problem prevention through proactive contracting. Scand. Stud. Law, Proactive Approach **49**, 21–34 (2006)
9. H. Haapio, *Next Generation Contracts: A Paradigm Shift* (Lexpert, Helsinki, 2013)
10. A. Hurmerinta-Haanpää, S. Viding, The functions of contracts in interorganizational relationships: a contract expert's perspective. J. Strateg. Contracting Negotiat. 1–21 (2019)
11. J. Kujala, S. Nysten-Haarala, J. Nuottila, Flexible contracting in project business. Int. J. Manag. Proj. Bus. **8**(1), 92–106 (2016)
12. P. Lahdenperä, A longitudinal view of adopting project alliancing: case Finland, in *10th Nordic Conference on Construction Economics and Organization, Emerald Reach Proceedings Series, Vol. 2,* Emerald Publishing Limited, ed. by I. Lill, E. Witt, pp. 129–136 (2019)
13. P. Lahdenperä, Making sense of the multi-party contractual arrangements of project partnering, project alliancing and integrated project delivery. Constr. Manag. Econ. **30**(1), 57–79 (2012)
14. S. Macaulay, Non-contractual relations in business: a preliminary study. Am. Sociol. Rev. **28**(1), 55–67 (1963)
15. J. Nuottila, Dissertation, Flexibility in agile contracts: Contracting practices and organisational arrangements. Acta Universitatis Ouluensis C Technica 726 (2019)
16. S. Nysten-Haarala, *Long-Term Contract: Contract Law and Contracting* (Finnish Lawyers' Publishing, Helsinki, 1998)
17. S. Nysten-Haarala, Why does contract law not recognize life-cycle business? Mapping of challenges for future empirical research, in *Corporate Contracting Capabilities, Conference Proceedings and Other Writings, Vol. 21*, University of Joensuu Publications in Law, ed. by S. Nysten-Haarala, Joensuu (2008)
18. S. Nysten-Haarala, J. Kujala, J.K. Aaltonen, Ketterät menetelmät julkisissa hankinnoissa. Liikejuridiikka **2**, 54–82 (2020)
19. S. Nysten-Haarala, N. Lee, J. Lehto, Hard and soft contracting, the human side of project business, in *IPMA World Congress Helsinki*, ed. by K. Kähkönen, A. S. Kazi, M. Rekola, 205–220 (2009)
20. Organisation for Economic Co-operation and Development Nuclear Energy Agency, *Nuclear New Build: Insights into Financing and Project Management*, OECD NEA No. 7195, Organisation for Economic Co-operation and Development, Nuclear Energy Agency, 2015
21. L. Poppo, T. Zenger, Do formal contracts and relational governance function as substitutes or complements? Strateg. Manag. J. **23**, 707–725 (2002)
22. J. Simonsen, A concern for engaged scholarship: the challenges for action research projects. Scand. J. Inf. Syst. **21**(2), 111–128 (2009)
23. A.H. Van de Ven, *Engaged Scholarship: A Guide for Organizational and Social Research* (Oxford University Press on Demand, Oxford, 2007)
24. A. van Marrewijk, N. Dessing, Negotiating reciprocal relationships: practices of engaged scholarship in project studies. Int. J. Project Manage. **37**, 884–895 (2019)
25. World Nuclear Association, *Lesson-learning in Nuclear Construction Projects, April, Report No. 2018/002*, https://world-nuclear.org/getattachment/e9c28f2a-a335-48a8-aa4f-525471a67 95a/REPORT-Lesson-learning-in-Nuclear-Construction.pdf.aspx (2018)

Chapter 6
Contracting Qualities that Challenge Reliability: A Case of the Utility Sector

Léon L. olde Scholtenhuis

Abstract This study uses the utility construction sector as a case to build the argument that specialisation, transience and price competition impede the reliable functioning of supply chains. These three contracting qualities obstruct the establishment of antecedents of mindfulness and the adherence to mindful organising principles. We offer three solution directions to improve contracting practice.

Keywords Reliability · Specialisation · Transience · Price competition · Utility sector

6.1 Introduction

Reliability-seeking organisations continuously put effort into avoiding allowing operational processes to lead to errors and eventually culminate in accidents. Studies of organisations in high-hazard industries, such as on aircraft carriers and in nuclear plants, demonstrate that reliability-seeking occurs in firms with organisational mindfulness and mindful organising practices. Reliability studies have found these characteristics in stable units (e.g. teams and departments) with clear boundaries that define who are internal and external to the organisation. This literature posits various antecedents of a mindful practice: leadership styles that endorse mindful organising, structures that manage the effect of organisational size on fragmentation [15] and the absence of extreme production pressures [20, p. 44].

Contracting supply chains are not as permanent and stable as the classic reliability-seeking organisations. Their outsourcing and contracting practices (here referred to as *contracting qualities*) challenge the realisation of the mindfulness antecedents. Construction is a sector where outsourcing is particularly common. Three typical contracting qualities of this sector are specialisation [3, 4], transience [13] and price competition [21, pp. 106–108]. For example, in the utility construction subsector, long supply chains of network owners and contractors concurrently work on a site

L. L. olde Scholtenhuis (✉)
University of Twente, Enschede, The Netherlands
e-mail: l.l.oldescholtenhuis@utwente.nl

© The Author(s) 2022
J. Hayes and S. Tillement (eds.), *Contracting and Safety*,
SpringerBriefs in Safety Management,
https://doi.org/10.1007/978-3-030-89792-5_6

to construct co-located, buried utility networks. Production pressures often push the performance of such projects to their limits. Based on this example case, we illustrate that contracting qualities can reduce mindfulness.

The remainder of this chapter introduces the concepts of organisational mindfulness and mindful organising. It then defines the case chosen and the utility sector and analyses how three contracting qualities challenge its reliability. We conclude by recommending improvements to existing contracting practices.

6.2   Organising for Process Reliability

Safety incidents, failures and other unexpected events can damage an organisation's health. Reliability-seeking processes aim to effectively cope with these unwanted and unanticipated events and their effect on performance [5, p. 51]. These processes occur on both the operational and strategic levels: Vogus and Sutcliffe [15] argue that, on the strategic level, *organisational mindfulness* arises when higher management shapes enduring practices, structures and cultures that favour mindful ways of thinking and organising. This, in turn, enables the dynamic, continuous, bottom-up process of *mindful organising* by frontline operational-level workers. Weick, Sutcliffe and Obstfeld [18–20] have defined five mindful organising principles that can help organisations identify details concerning potential threats to reliability and cope effectively with emerging unwanted events.

First, *preoccupation with failure* involves organisations treating any failure as an indicator of questionable system health. That is, they aim to learn from regular and thorough analysis of inconsistencies and near-errors. Reporting of errors then becomes more important than blaming individuals for their involvement in them. Second, *reluctance to simplify* contributes to the development of a comprehensive, rather than narrow and simplistic, interpretation of the current situation. Simplifications limit the identification of possible future operational scenarios, restrict precautions that people take against them and may lead to unintended negative consequences. To avoid this, organisations try to 'sense the complexity of the environment' by encouraging a diversity of perspectives, valuing scepticism and challenging assumptions about reality. Third, *sensitivity to operations* refers to the cognitive process that workers accomplish together by continuously developing and updating a collective understanding of evolving operational situations. This allows organisations to detect anomalies and 'catch errors in the moment'. Fourth, *commitment to resilience* reflects organisations accepting that errors will occur due to human mistakes, narrow specialties and complex technology [14]. Consequently, they develop abilities to cope with unanticipated surprises and then re-establish organisational processes. This involves improvising, utilising an individual's knowledge networks and deploying additional technical resources. The fifth principle is *deference to expertise*. With this, top managers acknowledge that expertise is more important than the decision hierarchy when problems in operational processes emerge. This leads to a collective and cultural belief that capabilities to resolve a problem lie in the system and that decision rights need to be given to those (frontline workers) with expertise on the event that has occurred [20].

The principles of mindful organising originate from case studies of stable and well-defined organisations such as aircraft carriers [17] and nuclear power plants [11] and mainstream organisations such as business schools [10] and hospitals [19]. All these studies have in common that they focus on mindful behaviour within an integrated organisational unit that is defined by clear boundaries. The lack of this structural characteristic challenges reliable performance. Vogus and Sutcliffe [15] have argued that the lack of structures through which leaders can share perceptions about the importance of mindfulness across different organisational levels challenges mindfulness. They also argue that a lack of practices to maintain a collective belief among organisational members that mindfulness organising is highly relevant exposes large organisations to fragmentation and lower mindfulness levels. Further, reliability is challenged in organisations with loosely coupled relationships [16] and in those that have functions with low task interdependencies [12]. Production pressure [20] may also overload cognitive tasks and reduce judgement and performance, thereby lowering sensitivity to operations. We now elaborate on these issues as they relate to the context of utility construction.

6.3 Contracting Qualities and Reliability in Utility Construction

The construction industry is characterised by project-based working [4] and a large degree of job specialisation and fragmentation [3, 4]. Along a project's lifecycle, the construction supply chain involves many stakeholders for different coordination, engineering, surveying and construction tasks.

The utility construction subsector builds and maintains networks such as gas and water pipes as well as electricity and telecommunication cables. It amounts to an extreme case of stakeholder fragmentation: most utilities in urban spaces are co-located in shallow trenches below pavements, under roads and along other rights-of-way. This creates physical interdependencies [9] since the relocation of one utility type often necessitates the replacement of other utility lines that are deployed close by. Further, the privatisation of utility networks has increased the number of stakeholders that are involved throughout their lifecycles. While their networks lie together in public space, network owners use distinctive strategies to plan, engineer and execute their own construction work. In addition, they outsource most construction work to a range of specialised contractors. This shapes a multi-stakeholder network involving several distinctive supply chains. This fragmented contracting context is impacted by the reliability-decreasing effects of contracting practices.

Currently, suboptimal alignment of stakeholders in utility streetworks causes network damages, deterioration of infrastructure, delays and project cost overruns. The societal costs of utility streetworks in the UK, for example, amount to 5.1 billion

GBP annually [2]. We now further illustrate the challenges to reliability based on the contracting qualities of specialisation, transience and price competition.

6.3.1 Specialisation

Specialisation moves tasks from within an organisation's hierarchy to the market. It allows vertically disintegrated organisations to focus on their core tasks while mobilising specialists as a flexible resource. For example, utility contractors hire subcontractors for activities such as excavation, welding, gluing cable joints, pipeline inspection, paving and surveying. Such specialisation diffuses the responsibility for the management of risks.

Risk shifting diffuses the coordination of risk along the contracting chain and effectively puts pressure on the risk management capabilities of field personnel [7]. When multiple contracting chains concurrently execute physically interdependent construction activities onsite, they report to distinct clients through different supply chains. This situation requires a formal main principal that is responsible for the coordination of all the organisations that are part of the distinctive supply chains. This coordination *between* supply chains is, however, missing. It makes risk management even more dispersed and uncoordinated.

Further, the involvement of multiple clients, contractors, engineers and trade specialists over time shapes a fragmented, complex and dynamic network where different organisations independently execute tasks. This limits their ability to collectively *sensitise themselves to how their operations interrelate*. Partial assessments of complexities, task interfaces, site risks and diffused decision-making power make it a challenging task for the supply chain to collectively develop a coherent view of existing risks and to develop risk mitigation capabilities.

In utility project supply chains, specialised field personnel will be involved on a site only temporarily and for a limited period. Excavator operators, welders and job supervisors move—often on a tight schedule—between different construction sites. In terms of *commitment to resilience*, this means that projects have limited flexibility in mobilising resources. Consequently, process interruptions may often not be managed directly and effectively. For example, when the discovery of polluted soil requires alternative digging methods (such as vacuum excavation), the required resources cannot be mobilised instantly because the multiple supply chains need to collectively decide on their mobilisation and allocate their costs.

Specialisation further means that network owners and their contractors work within a strict decision hierarchy. Consequently, contractors need formal approval to deviate from the original project plans. This rigid scope creates a tight decision hierarchy that places decision-making responsibility over all operational issues with clients that are distanced from the actual work on the construction site. This challenges the principle of *deference to expertise* since the specialist contractors and personnel onsite, who have the experience and knowledge to foresee and contain emerging problems (such as cable strikes, additional work due to polluted soil or leakages), need to obtain permission from higher up in their hierarchy to deviate from the detailed scope provided, even in the face of an emerging complex event. This process is further complicated since this approval might be needed from the multiple clients in the distinctive supply chains that concurrently work onsite. Overall, the tightly defined scopes reduce swiftness of incident responses.

6.3.2   Transience

Transience influences the reliability of utility projects in two ways. First, the members of stakeholder coalitions often change between successive construction projects [6, p. 3]. Main contractors will often mobilise different crews on different projects. They will outsource work lower down the supply chain by hiring freelance workers and small subcontractors on a project-by-project basis. Furthermore, stakeholders have different backgrounds, experience and expectations [13]. In these diverse and changing constellations, it is a continuous and time-consuming effort for a crew to *sensitise* themselves *to the complex operational interfaces* of their work with the onsite processes of other individuals.

Second, the transient and location-specific nature of streetworks means that contracted field personnel move between project sites. While working on a utility construction site, field workers establish links with other stakeholders, create knowledge networks and gain specific insights into site conditions within an area (such as about unmapped utility locations, polluted soil locations and previously unknown buried objects). This understanding of the local system contributes to the *avoidance of simplifications* regarding the project's reality. However, transience makes it difficult to maintain well-informed and updated knowledge about underground conditions at all the sites where a crew is working. In the Netherlands, for example, network owners hire contractors on a project-by-project basis. This creates transience: crews move between construction sites frequently, and once contractors move to new geographic regions, to a new project, their local knowledge evaporates. Moreover, the costs incurred make it unattractive for (sub-)contractors to fully explore project site complexities and build knowledge networks for each project. Transience thus reduces the return on investment in this knowledge development and reduces the contractor's ability to *commit to resilience*.

6.3.3   Price Competition

Utility contractors compete for work in tendering processes where network owners evaluate bids based largely on the tender prices [21]. Competitive tendering creates pressure on contractors to minimise their bid price. As a result, they may underestimate project complexity and offer low prices to win bids [21]. When contractors accept low-profit margins, they are under pressure to rapidly complete work to minimise financial losses. This, in turn, can inhibit any *reluctance to simplify* since contractors have less budget available to spend on analysing local construction site complexities and obstacles. Specifically, it exerts pressure to bypass error-anticipation activities such as mapping involved stakeholders, verifying utility maps, detecting interference with existing utilities and assessing how external factors (such as weather) might impact the project schedule.

Simplification is further incentivised by the productivity pricing method that contractors use to integrate production and risk coordination costs into a single price-per-metre figure. In the telecommunication sector, for example, contractors may cost work based on a length unit of cable installed or the number of houses connected to a network. This price includes direct construction costs and the additional costs involved in careful excavation work and damage avoidance (e.g. the costs for trial trenching, utility detection and vacuum excavation).

By using a single fixed price per metre of utility deployed, contractors essentially reduce the range of risks that can occur across varying project conditions (e.g. rural, residential, inner-city projects) to one 'standard' risk situation and price. Further, the pressures that result from productivity pricing can also reduce *commitments to resilience* since stakeholders who find themselves on projects that are more complex than 'the standard' have allocated too little time and budget to develop the knowledge and resource capacity required for a flexible and adequate response to emerging incidents.

Production pressures also influence the effectiveness of the health and safety regulations that supply chains use to avoid incidents and low-quality work. When under pressure, field workers seem to make trade-offs by deciding where they should adhere to regulations and where they can cut corners. Sometimes, for example, excavator operators cause minor damage by scratching a cable coating. Although the consequences of this damage are not immediately apparent, rules prescribe that workers should report this error to network owners and authorities. Instead, however, they often 'repair' the cables themselves. In this way, they risk that the network will be damaged further or need unexpected repairs in the future. Both physically and metaphorically burying mistakes, and moving on, thus lead to practices that hamper learning from failure: the *preoccupation with failure* and how to reduce this is reduced. Unequal power relations between the crew, subcontractors and upstream supply chain parties magnify this dynamic since these reduce the likelihood that crew will voice concern or criticise corner-cutting [7].

Finally, competitive pricing impedes *sensitivity to operations* by encouraging stakeholders to mindlessly comply with the bare minimum that rules allow. For example, a rule may require contractors to dig test trenches to verify utility locations before they start full excavation. Although this rule stipulates *that* trenches should be dug, it leaves it open to the crew's judgement as to *where* and *how many* trenches should be dug. Under cost pressures, contractors may underestimate the required number of trenches. Although this does not violate rules, it makes contractors less sensitive to risks and complexity.

6.4 Unintended, Unanticipated Events Occur

Essentially, mindfulness antecedents [15, 20] in supply chains are put under pressure by various contracting qualities. Specifically, specialisation complicates leadership and the establishment of shared perceptions concerning the importance of mindfulness between and within supply chains. In addition, the length of supply chains and the numerous interfaces between stakeholders result in incoherent and inflexible structures that do not create a mindful organisational setting. Third, although one might expect the physical co-location of cables and pipelines to create task interdependencies and necessitate mindfulness in order to coordinate risk, the loose coupling between the different supply chains and their inherent transience inhibits this. Fourth, production pressure and cognitive overload put further stresses on the judgement and performance of the supply chain stakeholders involved.

A logical consequence is that utility construction projects suffer from problematic coordination between and within their various contracting chains. Delays and overshooting budgets occur frequently. Tens of thousands of unexpected utility intrusions occur each year in the Netherlands alone [1], some causing injuries and fatalities. From informal conversations with contractors and networks, it also seems that they consider such damage as unavoidable. This suggests that errors have become unwanted but 'normal' by-products of utility construction work.

6.5 Recommendations

We offer three recommendations that may help the utility sector improve contractual conditions and deal better with unwanted events. First, integration mechanisms should be applied by outsourcing parties to address the consequences of specialisation and fragmentation. One way to achieve this would be that network owners in a geographical area (such as a street or district) jointly procure work on their co-located utilities. Joint procurement would introduce a single main coordinating contractor for all the supply chains involved. This would shape a clear line-of-command between contractors and subcontracted specialists and provide clarity over the responsibility for risk coordination. Another means to integrate supply chains is to mobilise a so-called boundary-spanning agent. In the Netherlands, these agents are called 'utility coordinators' [8] and have a dedicated task as informal liaison to create awareness of the complexities in collaborative streetworks.

Second, contractors could be incentivised to develop, maintain and share their local knowledge. By developing longer-term relationships, and by rewarding dedicated contractors that repetitively utilise the same crew in a specific region, network owners would reduce the likelihood that local knowledge fades away due to transience. Sharing utility location data in open databases would also contribute to this. Increased knowledge about local site conditions might also help contractors to make more realistic assumptions about existing project conditions and the feasibility of schedules and deadlines. This could also reduce production pressures.

Third, contract styles could be adapted to reduce the effect of price competition. One way would be to allow contractors to use pricing schemes that differentiate between projects that have different complexity levels. Complex projects could then involve higher rates to cover risk anticipation and containment than those in more straightforward projects. Another way could be to develop contracts that split the current tender price into a competitive element that includes only construction costs for the utility lines, plus a component for mindfulness-enhancing activities. Treating these aspects separately could reduce the incentives for contractors to cut costs on mindfulness-enhancing activities.

Finally, we note that these directions for improvement should be interpreted with the understanding that further empirical validation is required. We would therefore encourage future research to study contracting and reliability across different types of supply chains, both within and beyond the utility sector.

6.6 Conclusions

Principles for organising mindfulness to improve organisational reliability are well established but challenged by typical qualities of organising work in a contracting environment. In the utility sector, we see that *specialisation* reduces the development of shared perspectives on the importance of mindfulness and the development of a rich understanding of a project's reality. Next, the *transience* of both crew and the work onsite further disincentivises supply chains from developing resources to anticipate and mitigate unwanted events. Further, *price competition* puts pressure on mindfulness-enhancing activities, reducing resilience. Recommendations are thus to improve overall reliability and so reduce dangerous and costly failures are to reduce interfaces between supply chain organisations through integration mechanisms; create contractual incentives that reduce transience; and separate direct construction costs from mindfulness-enhancing costs.

References

1. Agentschap Telecom, *Code Oranje Blijft van Kracht Voor Graafsector (News Release, 15 May 2018)*, https://www.agentschaptelecom.nl/onderwerpen/kabels-en-leidingen/nieuws/2018/mei/15/code-oranje-blijft-van-kracht-voor-graafsector. Accessed 7 Dec 2019
2. M.H. Burtwell, M. Evans, W. McMahon, *Minimising Street Works Disruption: The Real Costs of Street Works to the Utility Industry and Society* (UK Water Industry Research Limited, London, 2006)
3. R.G. Eccles, Bureaucratic versus craft administration: the relationship of market structure to the construction firm. Adm. Sci. Q. **26**(3), 449–469 (1981)

4. R. Fellows, A.M. Liu, Managing organizational interfaces in engineering construction projects: addressing fragmentation and boundary issues across multiple interfaces. Constr. Manag. Econ. **30**(8), 653–671 (2012)
5. E. Hollnagel, *Human Reliability Analysis: Context and Control* (Academic Press, London, 1993)
6. M. Loosemore, A. Dainty, H. Lingard, *Human Resource Management in Construction Projects: Strategic and Operational Approaches* (Taylor & Francis, London, 2003)
7. V. McDermott, J. Hayes, Risk shifting and disorganization in multi-tier contracting chains: the implications for public safety. Saf. Sci. **106**, 263–272 (2018)
8. L. L. olde Scholtenhuis, T. Hartmann, A. G. Dorée, Exploring networked project coordination in combined utility streetworks. Eng. Proj. Organ. J. **10**(1) (2021)
9. H. Osman, Coordination of urban infrastructure reconstruction projects. Struct. Infrastruct. Eng. **12**(1), 108–121 (2016)
10. J.L. Ray, L.T. Baker, D.A. Plowman, Organizational mindfulness in business schools. Acad. Manage. Learn. Educ. **10**(2), 188–203 (2011)
11. P.R. Schulman, The negotiated order of organizational reliability. Adm. Soc. **25**(3), 353–372 (1993)
12. J.B. Sexton, E.J. Thomas, R.L. Helmreich, Error, stress, and teamwork in medicine and aviation: cross sectional surveys. The BMJ **320**(7237), 745–749 (2000)
13. J. Sydow, T. Braun, Projects as temporary organizations: an agenda for further theorizing the interorganizational dimension. Int. J. Project Manage. **36**(1), 4–11 (2018)
14. D. Vaughan, *The Challenger Launch Decision: Risky Technology, Culture, and Deviance at NASA* (University of Chicago Press, Chicago, 1996)
15. T.J. Vogus, K.M. Sutcliffe, Organizational mindfulness and mindful organizing: a reconciliation and path forward. Acad. Manage. Learn. Educ. **11**(4), 722–735 (2012)
16. K.E. Weick, Educational organizations as loosely coupled systems. Adm. Sci. Q. **21**(1), 1–19 (1976)
17. K.E. Weick, K.H. Roberts, Collective mind in organizations: heedful interrelating on flight decks. Adm. Sci. Q. **38**(3), 357–381 (1993)
18. K.E. Weick, K.M. Sutcliffe, Mindfulness and the quality of organizational attention. Organ. Sci. **17**(4), 514–524 (2006)
19. K.E. Weick, K.M. Sutcliffe, *Managing the Unexpected: Resilient Performance in an Age of Uncertainty*, 2nd edn. (John Wiley & Sons, San Francisco, 2007)
20. K. E. Weick, K. M. Sutcliffe, D. Obstfeld, Organizing for high reliability: processes of collective mindfulness, in *Research in Organizational Behavior*, vol. 1 (Jai Press, Stanford, 1999), pp. 81–123
21. G. M. Winch, *Managing Construction Projects: An Information Processing Approach*, 2nd edn. (Wiley Blackwell, Ames, 2010)

Chapter 7
Managing Workplace Safety in the Temporary Organisation—Theoretical and Practical Challenges Associated with Large Construction Projects

Heidi Helledal Griegel and Kenneth Pettersen Gould

Abstract Compared to permanent organisations, temporary organising causes different challenges for safety and learning at the workplace. We discuss how these challenges faced by project organisations are not sufficiently acknowledged or managed, either within theories of workplace safety or current safety management approaches in the construction industry. In addition, the chapter's insights contribute to an action-based approach to workplace safety.

Keywords Inter-organisational projects · Temporary organising · Workplace safety · Construction industry · Action-based approach

7.1 Introduction

In the construction industry, much work is organised following commercial tendering and contracting, creating temporary organisations for the delivery of single projects, with contract-based and temporary employment, dynamic plans, a competitive environment and constrained finances [7]. Temporary organising (through intra-organisational, inter-organisational or project-based organisations or firms [3]) has become common practice in many industries [1]. For example, construction within the energy sector is characterised by time-limited and inter-organisational projects. As such, workplace safety needs to be organised in ways that are flexible and meet time schedules. It also involves workers having to coordinate within and across newly established teams, make swift adaptations to particularities of the site and continuously adjust the way tasks are completed. Although organisational theory has been concerned with temporary organising for quite some time [10], few aspects of it have

H. H. Griegel (✉)
Statnett SF, Sandnes, Norway
e-mail: heidi.helledal@statnett.no

K. P. Gould
University of Stavanger, Stavanger, Norway

been applied in the research on workplace safety. In this chapter, we address how developments and challenges associated with organising work in inter-organisational projects have consequences for workplace safety. We build on organisational theory of temporary organising, as well as empirical research from construction sites in the Norwegian energy sector, to reflect on the challenges for workplace safety and safety management strategies.

7.2 Action-Based Theory of Temporary Organising

Lundin and Söderholm [10] emphasised the need for a theory of temporary organising some time ago. They elaborated on how organisational theory has gone through a development, from focusing on "decision-making" to focusing on "action", meaning that, while past organisational theories have considered control or decision-making to be at the core of the organisation [10], they stated that theory of temporary organising should focus on actions. We argue that an action-based approach is also helpful when analysing safety in inter-organisational projects organisations. Among other reasons, this is because the temporariness of these organisations leads to complexity, uncertainty, ambiguity [3] and less stable structures. This makes it difficult to plan; thus, it is important to understand the impact that practice and project processes have on safety, rather than the impact of pre-established decision-making structures [1] that may exist in permanent organisations. Four basic concepts that may influence and define actions have been presented, in order to understand temporary organisations [10]: time (in temporary organisations, time is limited); task (the reason for the creation of the temporary organisation); team (in the temporary organisation, the team is formed around the time available and the task to be performed); transition (temporary organisations are formed with the intention of achieving change: before and after the temporary organisation). Going forward, we will use this theoretical action-based approach to gain insight into workplace safety in construction projects.

The chapter is structured as follows: after first describing briefly how safety is conceptualised in current construction projects (based on empirical results from an ongoing study of safety management in construction), we focus on how safety can be understood in practice, using the four concepts of time, task, team and transition to describe workplace safety in temporary organisations.

7.3 Current Safety Management Approaches in Large Construction Projects and Some Limitations

Inter-organisational projects in the Norwegian energy sector normally consist of a construction client, one or two main contractors, as well as several sub-contractors and sub-sub-contractors. The contractors and sub-contractors may be large, medium or small companies, with sole proprietorships frequent for sub- and sub-sub-contractors. The time spent by the different contractors on a project may vary greatly. In an ongoing action research project, we are studying safety management in construction projects within the Norwegian energy sector, by conducting interviews (46 informants, both on-site managers and on-site workers), workshops and field studies. We have used this empirical study as the basis for discussing how project organising may influence workplace safety.

From an empirical point of view, we see that many of the safety management measures in the projects we are researching focus on complying with rules, set by regulators, the client or main contractor, with fewer measures contributing to actual risk awareness and safe practices in the workplace. Measures aimed at creating a clear separation between the client's and the contractor's responsibilities for risk and safety management (regulations, illustrations, etc.) end up creating "distance" between responsible actors that more often should be interacting and sharing safety information. The adding of coordinators and controllers creates new interfaces that increase complexity. In addition, too many procedures may impact the workers' flexibility and awareness. While it is not surprising that many coordination meetings are held in projects characterised by change and time limitations, the meetings are dominated by an instrumental rationality that safety can be controlled, given that responsibilities are clearly assigned and that workers are compliant with the decisions and plans made. We see that current safety management measures do not seem to be based on, or take sufficiently into account, the properties of temporary organising and the associated challenges for workplace safety. We question whether, by implementing many formalised procedures, coordinator roles, meetings and lines of safety responsibilities in the projects, safety management has trapped safety into rules and compliance [2] and as such forgotten about the relationship between actual practice and safety outcomes. Going forward, we will use the four concepts of temporary organising to describe how safety can be understood in practice, and, where appropriate, we use some examples to relate this approach to workplace safety in the inter-organisational construction projects that we are studying.

7.4 Safety and the Temporary Organising of Construction Projects

Issues such as time limitations or an overly task-oriented focus may of course also challenge workplace safety in more permanent and stable organisations. What we address here, however, is that these four concepts (time, task, team and transition) refer to issues that are not only present occasionally. In temporary organisations, like large inter-organisational construction projects, the four concepts are interrelated and refer to basic features of the project organisations. To illustrate, the energy sector construction projects we are studying consist of building new, or restoring existing, transformation stations and grid lines. Here, the grid line operators (i.e. the construction clients) and regulators set the predefined timeframe for the construction work. The work is divided into separate tasks, each with different and fixed deadlines; if a task is delayed, this has consequences for time. Tasks are executed by different specialised contractors or sub-contractors who operate in teams, sometimes created across company borders.

7.4.1 Time in Large Construction Projects

In permanent organisations, the objective is to succeed indefinitely; time is seen almost as eternal [10], and success—also in the domain of safety—is considered in light of contributions to long-term results [3]. Here, safety management systems are often developed to take care of occupational safety and company reputation, control unwanted events, manage compliance and support the continuous development of routines, procedures and checklists. Consequently, the improvement and management of safety is central for organisational survival [9]. In temporary project organisations, however, the end-time is decided before starting the project. In the construction projects we have studied, time is always a limitation, and all work operations and tasks may have their own deadlines. As in other temporary organisations [3], completing work on time is considered a key part of defining the project work as successful. It is known from research that limited time and deadlines for completion can make workers feel pressured to take unnecessary risks [7] and thus negatively influence workplace safety. One project manager (from a construction client) described to us how time limitations in this competitive industry influence safety, noting that doing work quickly poses a specific risk.

Despite the similarities to issues of time pressure and safety in permanent organisations, such findings do not sufficiently cover the intentional temporariness of project organisations. Drawing on a definition of a temporary organisation as "a temporally bounded group of interdependent organisational actors, formed to complete a complex task" [3], we consider time a fundamental issue of work, where the predefined ending of the organisation influences the organisational structure further, by leading to overlapping activities, many different professions on site at the same time, etc. The time constraints of each job are normally pre-planned and interlinked, so, if the tasks of one organisation become delayed, other tasks and organisations will also be influenced. One site manager (from a contractor) described the nature of multiple trades and multiple activities that occur on site simultaneously in order to meet deadlines. In his view, this creates a tension which can lead to accidents.

This description also indicates how different actors performing separate jobs at the same time leads to an increasing need for—but also challenges—coordination processes in the workplace. Others have previously emphasised how time-limitations, requirements for on-time completion [3] and many diverse individuals present at the same time lead to a focus on tasks rather than the "big picture". This will be further elaborated on below.

7.4.2 Tasks in Large Construction Projects

Construction projects can be said to be formed due to the need to perform specific tasks. As in other temporary organisations [10], construction projects within the energy sector are affected by increased specialisation, and many tasks are performed with varying degrees of uniqueness and connection to other tasks in the project. In current approaches to safety management, shared system safety knowledge and practices are usually viewed as important, and commitment to mutual and common organisational goals in general is recommended [11]. However, one on-site manager (from a contractor) described how they do not really value knowing too much about the overall project and safety goals when projects become large and temporary. For him, valuable knowledge is worker- and task-specific, and too much general project information can become too theoretical and irrelevant for the task at hand.

This understanding also depicts how these introductory safety courses are perceived as being too theoretical and general, as if they are not grounded in practice; it seems to highlight that the interdependencies between the tasks of the different professions are not really addressed. The task-oriented focus can make it difficult to create a shared concern for workplace safety. This is understandable, since safety goals are externally imposed [3] by the construction client or main contractor while the success of each actor is heavily influenced by individual performance and the completion of specific tasks. Sub-contractors are often paid per task [7], with each task having a deadline. We can therefore speculate as to whether this provides incentives for workers to skip or devalue safety-related activities that are not limited to the completion of their specific tasks. Also, some of the actors involved work on the project sites for less than one hour in total to complete their tasks and then leave for another project. So, even if each worker wants to perform their own task safely, it is understandable that it is challenging to familiarise themselves with the site and the project organisation and to grasp the totality of how their actions may have consequences for others and, as such, also influence workplace safety. It seems that organising work through tasks leads to a fragmented project organisation, and therefore, also a changing network of individual or team-based "pictures" of risk that others may not be familiar with.

7.4.3 Teams in Large Construction Projects

According to Gherardi [5], safety in organisations is embedded in common values and shared norms. It is a collective competence, emerging through collaborative practice, by being socially constructed, developed and transmitted to new members of the community [5]. In permanent organisations, work relations and cooperation can be long-lasting, and a random group of workers can form teams [10], surrounded by a stable environment and a feeling of a joint community and social relationships. Temporary organisations, on the contrary, exist as many separate teams that are created based on the completion of predefined tasks. Since the workers often belong to other (permanent) home-organisations besides the temporary project organisation [10], tensions [3] of belonging can occur, such as tensions between belonging to the current role in the home-organisation, with its current safety rules and social relationships, versus belonging to the teams in the coordinated, managed and controlled task-environment of the project organisation. In addition, the way safety management theory, based on permanent organisations, has emphasised the importance of collaborative community, commitment building [10] and mutual trust [3] is challenged by how work is organised in separate teams in temporary construction projects—particularly as the teams may contain members from the same or different companies and change by the hour. Each worker brings their own experiences and task-relevant competence [3] into the team, with participation in the teams being time-limited and focused on task completion. As such, across (and sometimes within) the teams in inter-organisational construction projects, there seems to be minimal shared safety knowledge and little pre-established trust. Combined with the focus on tasks, this creates project sites where project management focuses on project structure, roles and the need for temporal coordination rather than enduring social relations, trust and commitment building for workplace safety. In the projects we have studied, we have seen that many tasks are organised at the team level, and that the team level is important for safety. To exemplify, one of the operative workers in the projects we studied emphasised that, on first arrival at an ongoing construction site, they almost always team up with someone who has experience of the site until you "get the hang of it". The same worker also reflected on how such team relations are important for learning, emphasising the importance of social bonds between team members.

It becomes apparent that these teams are important for safety and learning; however, the fact that members belong to different organisations, the lack of team stability and the limited time that individual workers are part of the project organisation are all challenges for the long-term competence development of workers, as well as the general human resource management of the project team members [3]. Together with tensions between organisational cultures, social relationships and role emphasis, these are challenges to exchanging safety knowledge in the workplace and the sharing of information across company borders, project sites and hierarchical levels.

7.4.4 Transition in Large Construction Projects

We emphasise three aspects of transition that are important for understanding workplace safety in inter-organisational construction projects. The first concerns how transition in one sense mirrors the actual definition of a temporary organisation. The main goal for creating a temporary organisation is that, as a result of tasks, something should be changed or transformed. In our case, this aspect of transition concerns the establishment of temporary project organisations for the building of grid lines and transformation stations—where each individual task leads to transitions of varied significance and meaning. This aspect of transition is also the one that project management seems most attentive towards, and countless measures are implemented, such as meetings, coordinator roles and plans, to ensure timely progress to reach the desired transition.

The second aspect of transition concerns the frequent changes [8] occurring during the execution of tasks. Inter-organisational project sites are dependent on permits, technologies, equipment, weather conditions, etc., and task completion is influenced by a great number of plans, laws and regulations. This is accompanied by different degrees of uncertainty, goal ambiguity and complexity [3] related to each task. This significantly influences and challenges how changes can and should be dealt with while assuring workplace safety. During the construction phase in the projects we are studying, changes need to be continuously dealt with by the workers on site, and, as such, safety management relies heavily on the skills and actions of each worker. An operational risk assessment tool frequently applied to deal with changed preconditions is job safety analysis (JSA) [8]. However, like many of the existing "safety tools" applied in current construction projects, JSA was not originally developed based on inter-organisational principles [8] and temporary project work. We have found varying degrees of team members' participation in performing the JSA. Also, actors from different companies and teams, uninvolved in the particular job but who may have relevant information, are seldom included in the analysis [8].

A third important aspect of transition concerns the development and transfer of new experience, knowledge, skills and perceptions—technical, individual and social—among participants in the projects, as well as among participating organisations and between projects. For decades now, safety management theory, based on permanent organisations, has emphasised learning and knowledge transfer as crucial [6]. Although it is known that knowledge transfer and learning is a significant challenge [3] in construction projects [7], safety management theories [5, 6] developed based on permanent organisations are still frequently applied in research addressing the construction industry, without theoretically addressing temporality issues. Not surprisingly, this has given limited input to rethinking learning for workplace safety when the project organisation is large, complex and only temporary. Project management also seems to value sharing and developing knowledge concerning technical improvements; however, there seem to be fewer methods to bring these technical "learnings" to the next project. Generally, many of the objects or tools that are introduced in order to talk about, consider or address safety (e.g. risk assessments, coordination, meetings, etc.) are related to technical changes or a specific task, and we find fewer structures for the sharing of safety experiences and information related to organisational factors that are important for workplace safety within the projects. In some of the projects we have studied, project management performed evaluations after completing the work. However, many of the participating teams had already moved on to their next projects and those that had not wanted to. In addition, the technical focus emphasises that there is a "doing versus learning paradox" [3], as knowledge related to specific tasks, deliverability and completion seems to be more in focus than knowledge related to organisational factors and safety practices. These are issues we find relatable to how temporary organisations are embedded in a wider organisational context. Norwegian laws have tried to clarify the relationships, roles and responsibilities in construction projects [4], through e.g. defining separate responsibilities for the different project actors concerning risk management, and demanding coordinator roles, in both the planning and the execution of the projects. However, this has been with little or no improvement for practice-based learning and knowledge transfer, in our opinion, instead making safety management more about the separation of responsibility and coordination. Another Norwegian attempt to collaborate in an otherwise competitive commercial environment is the establishment of industry-wide networks [12] that have been established outside the project sites to exchange information, cooperate on improvement projects and create safety training courses.

7.5 Taking Account of Temporary Organising Towards an Action-Based Approach to Workplace Safety

The examples and suggestions we have provided in relation to managing workplace safety in temporary organisations together show that many of the traditionally stated preconditions for workplace safety, such as continuity of safety management for the continuous development of routines and supportive documentation (e.g. procedures), shared system safety skills and goals, enduring social relationships, developed trust, as well as continuous knowledge management and learning, are challenged by the way large construction projects can be conceptualised as temporary organisations [10]. A safety strategy focused on continuous development and long-term survival, often found and recommended for permanent organisations, is much more difficult—and perhaps even not possible in the same manner—in temporary projects. At a minimum, temporary organisations challenge basic assumptions regarding the role of time and the temporal limitations related to task, teams and transition that underlie dominant safety management theories. The intrinsic focus temporary organisations seem to have on tasks does not encourage understanding the "bigger picture" that working with safety often needs. We can also question whether part of the problem behind why the continuous development of social relationships and shared knowledge is so difficult is due to the team's different affiliations.

Although project organising has important advantages, such as flexibility [1], cognitive diversity [3], overcoming inertia [10] and increased specialisation opportunities, we suggest that, when approaching construction projects and other contemporary industrial environments, the safety field must take a step further back, rather than revisiting and adapting safety theories from studies of permanent organisations. We believe the role of temporality to be a gap in safety science worthy of further research and investigation. We hope to have demonstrated that the way that the predefined ending and the transformative nature of organising in temporary organisations affects the social processes involved in workplace safety is an important question for further research, as well as indicating that changes concerning safety thinking and approaches to workplace safety within temporary organisations, such as large construction projects, should be considered an important part of such research activities.

Ethical Statement   This work was approved by NORCE, The Norwegian Research Centre (Project no. 100315). Informed consent was obtained from participants and all data has been anonymised.

References

1. R.M. Bakker, R.J. DeFillippi, A. Schwab, J. Sydow, Temporary organizing: promises, processes, problems. Organ. Stud. **37**(12), 1703–1719 (2016)
2. M. Bourrier, C. Bieder, *Trapping Safety into Rules: How Desirable or Avoidable Is Proceduralization?* (Ashgate, Farnham, 2013)
3. C.M. Burke, M.J. Morley, On temporary organizations: a review, synthesis and research agenda. Human Relations **69**(6), 1235–1258 (2016)
4. Byggherreforskriften, Forskrift om sikkerhet, helse og arbeidsmiljø på bygge-eller anleggsplasser (2009)
5. S. Gherardi, A practice-based approach to safety as an emergent competence, in *Beyond Safety Training* (Springer Nature, Cham, 2018), p. 11–21
6. S. Gherardi, D. Nicolini, The organizational learning of safety in communities of practice. J. Manag. Inq. **9**(1), 7–18 (2000)
7. E.J. Harvey, Ph.D. Thesis, Loughborough University, 2018
8. H. Helledal, K.A.P. Gould, K.A. Holte, Job safety analysis in large construction projects—an inter-organizational approach to risk analysis and learning, in *European Safety and Reliability Conference, Venice* (2020)
9. Y. Li, F.W. Guldenmund, Safety management systems: a broad overview of the literature. Saf. Sci. **103**, 94–123 (2018)
10. R.A. Lundin, A. Söderholm, A theory of the temporary organization. Scand. J. Manag. **11**(4), 437–455 (1995)
11. V. Milch, K. Laumann, Sustaining safety across organizational boundaries: a qualitative study exploring how interorganizational complexity is managed on a petroleum-producing installation. Cogn. Technol. Work **20**(2), 179–204 (2018)
12. S. Winge, Ph.D. Thesis, Norwegian University of Science and Technology, 2019

Chapter 8
When the Project Ends and Operations Begin: Ensuring Safety During Commissioning Through Boundary Work

Anne Russel and Stéphanie Tillement

Abstract Ensuring safe performance in inter-organisational projects involves managing a whole range of organisational, occupational and spatio-temporal boundaries. Regarding future safety, the commissioning phase is crucial. Drawing from the case of the commissioning of a new nuclear installation, we highlight the challenges associated with the transition between the project and operations and show the socio-material and temporal arrangements that support or hinder boundary work.

Keywords Nuclear industry · Inter-organisational projects · Commissioning · Boundary work · Process safety

8.1 Introduction

Since the 2000s, the safety implications of outsourcing work activities have become the subject of scrutiny from safety authorities, politicians and civil society. Most studies on that topic highlight a negative link between safety and outsourcing [7, 11]. With few exceptions [4], research conducted in the process industries focuses on permanent organisations (often maintenance) and does not study "from the inside" the way outsourced activities are actually carried out. Yet, lots of outsourced activities are now organised and coordinated within temporary configurations, as so-called inter-organisational projects (IOP) [10]. There is a need to better understand the nature of interactions between the principal and the contractors in such organisational settings and how work is actually carried out and negotiated in relation to safety. Within IOPs, the design, construction and operation phases are generally carried out by actors from different companies. The transitions from one phase to another entail significant challenges for both project performance and future safety, as they require intense coordination between actors from various sectors and occupations. These key moments often disclose conflicts of interests, contractual disagreements and a need for continuous negotiation. Among them, commissioning is often considered

A. Russel (✉) · S. Tillement
IMT Atlantique, LEMNA, Nantes, France

© The Author(s) 2022
J. Hayes and S. Tillement (eds.), *Contracting and Safety*,
SpringerBriefs in Safety Management,
https://doi.org/10.1007/978-3-030-89792-5_8

highly critical. It is the final opportunity to test the operational feasibility and safety in future operation by identifying and fixing all remaining deficiencies and errors [6]. Successful commissioning involves managing a whole range of boundaries, i.e. organisational, occupational and temporal [15]. Several authors have highlighted the role of boundary work [3] supported by actors or objects, but rarely in the context of high-risk IOPs and in relation to safety. In addressing our theoretical concerns, we develop the following research question to guide our empirical study: *how and under which conditions can boundary work contribute to safe performance during the commissioning phase of projects?*

To answer the research question, we investigate the commissioning of a new facility at a nuclear waste storage site. This enables us to draw attention to the specific difficulties encountered by the actors in this transition phase but also the socio-technical arrangements that are negotiated throughout this phase to manage boundaries and overcome these difficulties. Finally, we discuss the important lessons at the organisational and contractual dimensions.

8.2 Outsourcing, Projects and Safety

In the literature, outsourcing has been described as "the practice where a public or private organisation contracts another organisation or individual—usually through a process of competitive tendering—to undertake specified tasks, such as cleaning, transport or maintenance or even provision of a product" [7, p. 284]. Over the last 30 years, these practices have developed in high-risk organisations, mainly for economic or strategic reasons. It was seen by industries as a way to reduce labour costs and often driven by a strategy that focused on the core competences and/or sought enhanced organisational flexibility.

Outsourcing has often been identified as a causal factor of accidents. Some studies have highlighted the adverse effects on process and occupational safety, by highlighting the associated socio-technical risks (e.g. loss of internal know-how and competences for the principal) and organisational risks (e.g. excessive dependency on subcontractors with rare competences). Very recently, researchers have made the link between outsourcing, inter-organisational complexity and safety more explicit. Outsourcing can increase economic pressure, disorganisation and dilution of abilities [5]. In a context of high competition, contracting companies may take safety shortcuts and transfer risks to the lower supplier in the supply chain [4]. But this does not have to happen; under certain organisational conditions, outsourcing may foster operational safety and reliability. For example, long-term organisational relationships can contribute to development and maintenance of good social interactions between the different companies and groups of workers involved [5].

In line with these works, our research aims to better qualify organisational and professional conditions that may affect safety in IOPs, where work is distributed between multiple organisations and disciplines and performed in temporary settings.

In the literature on complex projects, the links between performance, safety and interface management are a strong and recurrent issue [2, 9, 12]. The many boundaries (organisational, occupational, temporal, spatial) that generate several major problems in terms of communication and knowledge sharing [2, 14], role distribution and articulation work among various project phases and stakeholders or power and occupational jurisdictions [1]. Depending on the project and how it interacts with permanent organisations, the more problematic boundaries are not necessarily the ones between organisations but rather between occupational groups within a single organisation [12]. In enhancing coordination, many authors have highlighted the role played by specific individuals, known as boundary-spanners [14], who contribute to project performance through their ability to improve the sharing of information and knowledge [2] between the different organisations or professional groups within these projects. Boundary objects [8] appear to be just as important for transferring and translating knowledge in fragmented organisational contexts. They support ongoing negotiation between the various actors while acknowledging the specificities of each actor's activities, rhythms of work and skills [13, 16]. But, as shown by [1], artefacts may also reinforce boundaries and impede coordination when they are used to reassert authority and legitimacy over tasks.

We draw on this literature to investigate the case of a high-risk inter-organisational project carried out in a nuclear waste storage plant.

8.3 A Safety-Critical Project: Building and Operating a New Facility at NucStor Plant

Our study focuses on the 18-month commissioning phase of an ongoing IOP that aims at designing a new installation that will be dedicated to the reinforced control of nuclear waste packages. The plant in which the project takes place is operated by NucStor,[1] an organisation specialising in nuclear waste storage. As such, it is responsible for the safety of the storage facilities and supervises the whole range of activities at the site. But most of the latter, including production, is outsourced. Production, which consists of the reception, control, conditioning and storage of waste packages, is entrusted to WasteCorp, an external company specialising in the construction and decommissioning of nuclear facilities. The course of the project has followed the classical steps described in the literature, i.e. design, construction and commissioning, each being led by a specific actor. NucStor's headquarters led the design phase, the construction was managed by a contractor under the supervision of NucStor's project team on site, and the commissioning was performed conjointly by the project team and WasteCorp. Our study focuses on this last phase, which is highly critical, due to the many socio-technical and organisational interfaces that are brought into play. The new unit will be integrated into the existing production process and, in line with the current situation, operated by WasteCorp. Its incorporation into

---

[1] For confidentiality reasons, all company names are pseudonyms.

the existing socio-technical system is crucial for future safety. In completing this project, NucStor and its contractors are confronted directly with occupational and process safety issues since operating the facilities (both current and future) involves manipulating radioactive nuclear materials. Defining the future organisation of work requires cooperation and coordination between members of NucStor and contractors, notably during the commissioning phase. This requires articulating both current practices and knowledge, and those necessary to operate the future plant. In doing this, the tests performed during commissioning are crucial.

To understand how boundaries are managed to ensure safe performance, we draw on interviews with NucStor's project team, WasteCorp's operators and the project pilot.[2] We also observed tests and project meetings during the commissioning phase. Data were collected from September 2018 to July 2019 and followed by a qualitative analysis of interview transcripts and observation notes. Data coding was carried out manually and followed an iterative analysis process based on the comparison of field data with existing literature.

8.4   Challenges Encountered in the Commissioning Phase

As the future operator, WasteCorp, along with NucStor, plays a key role in the commissioning phase, which constitutes a test for the future operability and safety of the new installation. This involves facing emergent and often unexpected technical events that require specific technical and social skills to manage. From the very beginning of the commissioning phase, WasteCorp operators express difficulties in taking ownership of the facility and operating it as originally defined by the NucStor design team at headquarters.

Example of Unexpected Problems Encountered During an Inventory Test

An inventory test is carried out on an insulating box containing several waste packages. As the operator tries to move the packages from the box to the sorting area, he faces two problems. Firstly, the remote-control arm is not easy to manipulate and the clamps cannot pick up parcels that are too small (Fig. 8.1a). Secondly, when moving an oversized package, the package collides with the wall that separates the box from the sorting area, because the ceiling is too low (Fig. 8.1b).

---

[2] The project pilot is employed by WasteCorp. Her role is to monitor the operators' commissioning work in the new installation and to support inter-organizational coordination between WasteCorp's operators and NucStor's project team.

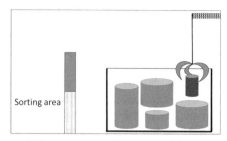 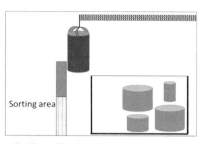

a) The remote control arm cannot pick up parcels that are too small

b) The ceiling is too low to allow the transport of large packages of waste in the sorting area

Fig. 8.1 Illustration of problems encountered during an inventory test

These problems arise from errors in the design and construction of the inventory room, which turns out to be too narrow. Alerts issued by operators regarding the handling of packages and the size of the room were not taken into account during design and construction. As a result, the operator has to enter the box to manually retrieve the parcels, which is theoretically forbidden.

Three main socio-technical issues are revealed, i.e. poor ergonomics of the installation, poor training of operators and poor communication between headquarters and the site, all having potential safety implications (Table 8.1).

Firstly, the operators complain that the work areas are too tight: once equipped with the regulatory protective suits, they lack adequate space to move around and properly handle the various tools and machines, which leads to additional handling

Table 8.1 Socio-technical problems and implications for safety

Main challenges	Potential safety implications
Ergonomics of the workplace	Imprecise handling of waste
	Contact with radioactive packages
	Handling time too long in contaminated areas
Training of operators	Partial knowledge of risks and preventive measures
	Lack of monitoring and traceability of control activities
Communication	Poor management of key interfaces for future process safety
	Little reactivity in the face of emergent technical issues
	Potential lack of transparency

time. Some tools are poorly designed and unusable as built. More importantly, this compels operators to develop deviant practices (some of which are theoretically prohibited) which leads to unexpected occupational and process safety issues.

Secondly, the operation of the new facility requires developing new skills associated with the new techniques and activities. The operators feel they are left to themselves in the process of acquiring these new skills and so must learn by trial and error on the job. Several reasons are advanced: lack of dedicated training times because of production pressures and insufficient support from hierarchy, principal and suppliers, non-recognition of the increased technical complexity and difficulties in complying with regulatory work (fulfilling the new operating procedures provided by design teams).

Finally, communication difficulties between headquarters and the site are revealed during commissioning. On the one hand, the design team is critical to the on-site project team and operators, who, according to designers, do not fully comply with their technical directives and requirements. Concretely, this is reflected in the imperative tone of their e-mails, their attitude of "principal" and their indifference to the technical constraints experienced by the operators in the field. On the other side, both the NucStor project team on site and WasteCorp operators express difficulties in communicating with headquarters members who are accused of having too abstract a vision of the installation, of not taking into consideration the technical feasibility of the current tools and of not visiting the site often enough. These attitudes, added to the geographic distance, tend to reinforce within NucStor the fault lines between the "people on the spot" and "the people at headquarters".

These problems have safety implications that can be observed in the case of projects involving workers belonging to the same company. But they tend to be reinforced in the face of outsourcing due to the multiplicity of interfaces.

In the face of these difficulties, NucStor and Wastecorp employees have conjointly and progressively developed and refined structural and operational local arrangements to manage interfaces and pursue their activities during the commissioning of the new installation.

8.5 Boundary Work for Improved Reliability

Early in the commissioning phase, the actors have engaged in intense boundary work, defined as all the socio-material arrangements and devices carried out to manage the many interfaces and associated risks during the commissioning phase. It relies on three key levers: individual, material and temporal.

At the beginning of the commissioning phase, a project team was set up on site to enhance integration and coordination between steps and stakeholders. It gathers NucStor representatives with specific key expertise and a project pilot from WasteCorp at the interface between the organisations. The analysis shows that the project pilot acts as a boundary spanner between the NucStor project team and the WasteCorp operators. Thanks to her "double hat", the project pilot benefits from the confidence of both companies, which enables her to easily communicate the instructions coming from the client and the requests and needs formulated by the operators. Her position supports inter-organisational coordination by allowing fluid and rapid interactions between the NucStor project team and WasteCorp operators. As contractors, WasteCorp operators must not receive direct instructions from the NucStor project team, so by integrating the project pilot with the project team, NucStor can also indirectly transmit instructions to operators, while avoiding perceptions of interference. As a trained engineer and manager of the operators in charge of commissioning, the project pilot also shares a professional experience on the ground with the WasteCorp operators, which is very valuable to the project team. During the project meetings, she is considered to be a fully fledged member capable of understanding the theoretical operation of the installation's different processes. At the same time, her daily involvement in the commissioning activity and her ability to provide permanent support for the operators' work enables her to develop strong expertise in the materiality of the installation and its constraints and to acquire professional legitimacy in the eyes of the operators.

Throughout the commissioning phase, the installation itself constitutes a boundary object between two occupational groups that each have different visions of the commissioning's times: NucStor engineers and WasteCorp production operators. While for the project team, commissioning is the end of the project; it represents for the operators the beginning of their new production activity. The materiality of the installation helps operators and engineers in managing and solving together the various problems encountered during commissioning (Box 2).

The Installation as a Boundary Object Between NucStor and WasteCorp

11:00am: During a technical test, the operator in charge of piloting the installation notices a design problem: a sensor is incorrectly positioned, blocking the operation of the monorail. He shares the problem with the project pilot who calls the NucStor maintenance manager to inform him of the problem.

3:30pm: The maintenance manager joins the operator on the installation. Together, they inspect the sensor problem: the operator explains that the sensor is malfunctioning because it sometimes rubs against the wall, blocking the operation of the monorail; the maintenance manager suggests unscrewing the sensor and positioning it in another place.

4:00pm: They agree that a modification must be made. The maintenance manager leaves the installation. He plans to make new supports for the sensors. The operator goes back to the control room and informs the project pilot of the decision. The project pilot integrates this modification request and modifies the test schedule accordingly.

By discussing the same object and confronting its materiality, NucStor project members and WasteCorp operators come to a better understanding of their respective problems. While each develops its own vision of the project and its constraints, the installation and the performed tests support the discussions on their respective roles regarding the commissioning and facilitate the cohesion of the different actors around the same objective. The installation helps in revealing the work carried out by the operators: by making their contribution in enacting the different processes visible to the engineers, the operators legitimise their role in the project and reaffirm their technical expertise vis-à-vis NucStor engineers.

Finally, our study reveals the positive contribution of the contractual relations established between NucStor and WasteCorp over almost 30 years due to the cooperation between both entities throughout the course of the project. These long-term relationships have allowed them to act jointly in the face of the many difficulties encountered during commissioning. The project has strengthened trust-based relationships between engineers and operators from both organisations engaged in the project. WasteCorp's operators have developed a sense of belonging to the project, which goes beyond their initial organisational affiliation. But paradoxically, the inter-organisational cohesion within the project has created fault lines and reinforced the boundaries between the local and global levels, partly due to a lack of devices and tools to support communication and coordination.

8.6   Discussion and Conclusion

Literature on IOPs describes the challenges in terms of knowledge sharing and coordination between distinct stakeholders, with a strong focus on inter-organisational boundaries. As a transition between project and operation, commissioning is viewed as a crucial step for the success of a project and future safe operations, involving temporal boundaries. In this chapter, we have discussed the challenges collectively faced by actors during this transition and the many boundaries involved. In particular, we highlighted how a lack of involvement of the future user during the design and construction phases undermined the ability of operators to quickly learn how to operate the new installation and manage commissioning. But we have also shown the role played by boundary-spanners and boundary objects in overcoming these obstacles and successfully completing the commissioning. These arrangements supported boundary work, which was particularly efficient at the inter-professional and inter-organisational levels. In our case, the existence of long-term contractual relations between client and contractor [9] also helped in managing these boundaries within the project, thanks to inter-individual relationships at the local level.

Finally, evidence from our research suggests that interfaces between the principal and its contractors are not necessarily the most problematic. When they are not considered, discussed and equipped throughout the project course, and notably in the final transition phase, intra-organisational and geographical boundaries prove to be just as crucial as inter-organisational ones for ensuring safety and performance of IOPs.

Acknowledgements This work was supported by the RESOH Chair. We are very grateful to NucStor and WasteCorp workers for their time and information.

Ethical Statement This work adhered to the research ethics that are stipulated in the "RESOH Chair convention" that complies with relevant legislation regarding ethical conduct of research. Informed consent was obtained from participants, and all data has been anonymised.

References

1. B.A. Bechky, Sharing meaning across occupational communities: the transformation of understanding on a production floor. Organ. Sci. **14**(3), 312–330 (2003)
2. P. Bosch-Sijtsema, L.-H. Henriksson, Managing projects with distributed and embedded knowledge through interactions. Int. J. Project Manage. **32**(8), 1432–1444 (2014)
3. A. Langley, K. Lindberg, B. Mørk, D. Nicolini, E. Raviola, L. Walter, Boundary work among groups, occupations and organizations: from cartography to process. Acad. Manag. Ann. **13**(2), 704–736 (2019)
4. V. McDermott, K. Henne, J. Hayes, Shifting risk to the frontline: case studies in different contract working environments. J. Risk Res. **21**(12), 1502–1516 (2018)
5. V. Milch, K. Laumann, Sustaining safety across organizational boundaries: a qualitative study exploring how interorganizational complexity is managed on a petroleum-producing installation. Cogn. Technol. Work **20**, 179–204 (2018)
6. P. Oedewald, N. Gotcheva, Safety culture and subcontractor network governance in a complex safety critical project. Reliab. Eng. Syst. Saf. **141**, 106–114 (2015)
7. M. Quinlan, I. Hampson, S. Gregson, Outsourcing and offshoring aircraft maintenance in the US: implications for safety. Saf. Sci. **57**, 283–292 (2013)
8. S.L. Star, J.R. Griesemer, Institutional ecology, translations and boundary objects: amateurs and professionals in Berkeley's Museum of Vertebrate Zoology, 1907–39. Soc. Stud. Sci. **19**(3), 387–420 (1989)
9. W. Starbuck, M. Farjoun, Organizing at and beyond the limits. Organ. Stud. **28**(4), 541–566 (2007)
10. S. Stjerne, J. Söderlund, D. Minbaeva, Crossing times: temporal boundary-spanning practices in interorganizational projects. Int. J. Project Manage. **37**(2), 347–365 (2019)
11. A. Thébaud-Mony, Travailler peut nuire gravement à votre santé: Sous-traitance des risques, mise en danger d'autrui, atteintes à la dignité, violences physiques et morales, cancers professionnels, La Découverte, 2016
12. S. Tillement, C. Cholez, T. Reverdy, Assessing organizational resilience: an interactionist approach. Management **12**(4), 230–264 (2009)
13. S. Tillement, J. Hayes, Maintenance schedules as boundary objects for improved organizational reliability. Cogn. Technol. Work **21**(3), 497–515 (2019)
14. M.L. Tushman, Special boundary roles in the innovation process. Adm. Sci. Q. **22**(4), 587–605 (1977)
15. J. Whyte, T. Nussbaum, Transition and temporalities: spanning temporal boundaries as projects end and operations begin. Proj. Manag. J. **51**(5), 505–521 (2020)
16. E.K. Yakura, Charting time: timelines as temporal boundary objects. Acad. Manag. J. **45**(5), 956–970 (2002)

Chapter 9
Outsourcing Risk Governance: Using Consultants to Deliver Regulatory Functions

Jan Hayes, Lynne Chester, and Dolruedee Kramnaimuang King

Abstract The Australian gas supply industry provides a case study of outsourcing by economic regulators. These regulatory agencies rely on the engagement of external consultants for technical expertise who are effectively tasked with finding ways to reduce proposed expenditure, even for safety-related items. Empirical evidence shows that economic regulators uncritically accept this advice. Such outsourcing raises the possibility of significant impact on technical regulatory outcomes.

Keywords Economic regulation · Gas industry · Expert consultants · Public safety · Risk management

9.1 Introduction

Since the early 1990s, there have been significant changes to the governance of utilities such as electricity and gas, communications and transport which were owned and operated by government. These previously vertically integrated monopolies have undergone restructuring with segments being privatised and subjected to competition. Where such segments have natural monopoly characteristics, they continue to be subject to government control in the form of economic regulation. Each sector is also subject to technical regulation to ensure that safety and quality standards are maintained. Both kinds of regulation have been influenced by another significant public policy change. Deregulation, or 'red tape reduction', has meant effectively a move from prescription towards goal-based regulation combined with forms of

J. Hayes (✉)
RMIT University, Melbourne, Australia
e-mail: jan.hayes2@rmit.edu.au

L. Chester
The University of Sydney, Sydney, Australia

D. K. King
Massey University, Palmerston North, New Zealand

J. Hayes and S. Tillement (eds.), *Contracting and Safety*,
SpringerBriefs in Safety Management,
https://doi.org/10.1007/978-3-030-89792-5_9

self-regulation governance. This has occurred across a range of sectors, including the technical and economic regulation of utilities.

Concurrently, an increased reliance on competitive market forces has resulted in a progressive rise in outsourcing and contracting-out across the private sector. This chapter considers one form of outsourcing adopted in the regulatory environment, that is, the use of contractors to perform assessments and evaluations to assist an economic regulator to perform its functions. We believe the use of contractors and consultants by regulatory agencies is not uncommon; yet it has received little academic attention despite the potentially significant implications. We address this issue by drawing on our empirical research regarding the impact on technical compliance, and thus safety, of the economic regulatory regime for the Australian gas pipeline and distribution network (gas supply) sector.

9.2 Effective Regulation and Effective Regulators

Proponents of a market-based approach to regulation emphasise that, in addition to state regulation, industries 'self-regulate' using industry standards, norms established by industry associations and similar. Ayres and Braithwaite [1] support the need for responsive regulation, i.e. to use a mixture of state regulation and self-regulation, and a range of strategies from encouragement to sanctions to ensure compliance. Policy regarding responsive regulation has been influential although it 'has been joined … by "risk-based" regulation' [2, p. 65]. In risk-based or goal-setting regulation, the regulations require that companies use a risk-based process to ensure that outcomes are adequately addressed, rather than prescribing detailed measures that companies must take.

The change in regulatory policy and frameworks has led to change in the day-to-day practices of regulatory agencies. Regulators are now required to assess such questions as 'how *well* the duty holder has carried out the risk management process and how *effective* its controls are' [8, p. 210]. They may also seek to influence duty holders to comply by interacting with a broad range of stakeholders including company executives, extending well beyond the skills of many traditional inspectors [4].

The subject of our empirical research, the Australian gas supply sector, is subject to goal-based technical regulation that seeks to ensure public risk—as a result of pipeline failure—is reduced to a level that is as low as reasonably practicable. Questions of regulator capacity are key to the integrity of outcomes.

9.3 Outsourcing of Regulatory Functions

The privatisation and deregulation agenda has seen many formerly government operations move to the private sector. Most commonly, these have been service delivery operations. Of interest here are a different set of government operations linked to regulatory oversight. In such cases, public good outcomes do not necessarily align with the commercial imperatives driving private sector providers.

The building construction sector in the UK provides one example of the possible link between privatised—outsourced—regulatory functions and public safety outcomes. Regulations address a range of requirements for the quality of building works. A key regulatory function is building inspection to check that work has been done to the requisite standard before permits are issued. Part privatisation of this function has raised concerns about the extent to which the actions of inspectors are driven by commercial pressure from clients, rather than a desire to enforce compliance with relevant standards [6].

The UK is not alone in this style of building control regulatory enforcement. Changes to privatise building control came about through a combination of both top down and bottom up pressures. Introducing private sector inspectors in several countries, including Australia, has been driven by government policy in the search for greater efficiency and effectiveness. One study of private sector involvement in building control in three Australian states found that it had 'resulted in more technical efficiency and effectiveness of regulatory enforcement' [9, p. 206]. Private sector involvement was seen to increase the skill level of inspectors overall. On the other hand, there are also negative effects. Private sector involvement is seen to have an adverse impact on accountability as conflicts of interest and weak auditing can result in lack of effective oversight. Similar results were found in building inspection practices in several Canadian states [10].

9.4 Economic and Technical Regulation of Gas Supply

Our case study focuses on the interaction of economic and safety/technical regulation. Utilities such as the supply of natural gas are considered natural monopolies. Economic theory assumes that the prices charged by a monopolist will be set at a level that generates excessive profit, and thus, there is a need for economic (price) regulation in the absence of the pressures of market competition.

Textbook theories of optimal pricing for regulated businesses assume that regulators are fully informed about costs, technology, and consumer demand and can therefore impose cost minimisation obligations on the regulated business. In reality, regulators have imperfect and less information than the regulated business. Consequently, issues of relevant experience and expertise on the part of economic regulators are critical as we will show.

A key consideration for economic regulation is the need to balance the interests of investors wishing to make an adequate return, with the interests of consumers who many be subject to exploitation in monopoly conditions [3]. This objective acts in tension with technical regulation to encourage expenditure on items such as maintenance and inspection to ensure public safety is maintained. In summary, the interaction between company and regulatory practices operates as follows:

1.   A need for physical system changes is identified by a regulated business in accordance with the risk management processes of technical legislation and standards.
2.   The business submits a safety case update to the technical regulator who assesses it inhouse and ultimately grants approval.
3.   The regulated business then submits a proposal to the economic regulator which contains risk-based arguments for expenditure to comply with its approved safety case.
4.   The economic regulator assesses the proposed safety expenditure using external consultant advice which interprets and applies technical regulatory requirements that commonly differ from that of the regulated business.
5.   The economic regulator does not consider a regulated business' compliance with a safety case to be tantamount to a regulatory obligation of the economic regulatory regime.
6.   The economic regulator determines, based on the advice of its external consultants, that a regulated business can spend less than it has proposed for safety purposes.
7.   The regulated business, in light of the economic regulator's decision for approved expenditure, reviews its asset management strategies which may include revision of asset risk profiles.
8.   Should a regulated business decide to spend more than that approved by the economic regulator for safety expenditure, there is no guarantee that this additional expenditure will be approved by the economic regulator.

These steps demonstrate the convoluted decision-making process regarding safety expenditure that the dual regulatory regimes impose on regulated businesses.

9.5   Method

Our empirical study collected qualitative information through semi-structured interviews, and conducted a document analysis, to gain an understanding of the economic and technical regulatory frameworks and how they operate in practice.

Forty-nine semi-structured interviews were conducted with senior employees of: the Australian Economic Regulator (AER); the Economic Regulation Authority of Western Australia (WAERA); the technical regulators from four Australian states (New South Wales, Victoria, South Australia and Western Australia); eight companies with majority ownership interests in six Australian gas transmission pipeline and eight distribution networks (in all Australian states and territories except Tasmania) subject to full economic regulation; peak industry bodies; and, current or former consultants to the gas sector.

The document analysis included the determinations of economic regulators, proposals submitted by companies to economic regulators, the performance reporting of technical regulators, and annual company reports (including financial statements).

We also examined in-depth five case studies from AER and WAERA decisions that directly related to proposed company expenditure for safety and asset integrity purposes.

9.6 Findings

Our analysis found that the technical and economic regulatory regimes are in significant tension [7]. The amount and timing of expenditure on projects linked to pipeline integrity is impacted by the decisions of the economic regulator. This is not to suggest that, despite decisions by the economic regulator, we found any regulated gas business would fail to spend money on something that they considered to be an immediate safety problem. The impact is far more subtle.

The safety of gas supply systems is 'a dynamic non-event'. For low frequency events such as major pipeline failures, the past is not a good predictor of future performance. It is easy to become complacent given that the warning signs of failure can be hard to spot. In this operating environment, significant investment in both time and money is required to ensure that the state of the system, including all risk controls, is well understood, and small problems are identified and monitored or fixed to prevent a catastrophe in the future. The technical regulatory regime is designed with this in mind but the economic regulatory regime takes a very different orientation towards all expenditure, even that which is safety-related.

One key factor uncovered by our research is the use of technical consultants by the economic regulator and the influence that consultant advice has on determinations. Many gas company participants expressed criticisms and frustrations about the consultants engaged by the economic regulator. Limited availability of expertise, poor choices, questionable independent advice, and conflicts of interest were consistent themes in the comments of company participants.

On the subject of expertise, company participants were particularly critical of the lack of specialist knowledge in the consultants engaged by the economic regulator, emphasising that consultants often had no relevant gas industry experience at all.

If the issue was simply lack of expertise, we might expect a variation in the assessment of company expenditure proposals with consultants suggesting either more or less expenditure is warranted. In fact, we found the proposed variation is always in one direction. The economic regulator's consultants are always of the view that the proposed expenditure is too high. According to some company interviewees, this reflects the implicit scope of consultants' work set by the economic regulator. In their view, the consultants are effectively paid to make the argument that the cost information provided by companies is inflated.

The consultants' task includes the provision of significant expert judgement, and company interviewees were of the view that consultants have a conflict of interest.

There are financial and reputational benefits in working for the economic regulator. As a result, consultant advice tends to support the economic regulator's position that companies overestimate necessary expenditure and so consultants typically 'conclude' that proposed expenditure is more than required to ensure system safety.

Gas supply companies also use external expertise, and this limits the consultant pool for the economic regulator due to conflicts of interest in that consultants cannot take on work reviewing regulatory submissions written by companies that are their clients. Both gas companies and regulators agree that this creates difficulties, effectively excluding many experts from working with the economic regulators due to conflict of interest issues.

Our study also included an in-depth review of five case studies of economic regulatory determinations which show the very influential role that external consultants play in forming the economic regulator's understanding of safety and its subsequent decisions. In all cases, the economic regulator accepted the expert consultant's assessment and recommendations.

In one case study, the regulated business proposed expenditure for modifications to the pipeline system to allow comprehensive integrity testing. The AER's consultant concluded such testing was not required and negotiation between the economic regulator and the regulated business moved to known failure mechanisms versus unknown failure mechanisms that inspection is designed to find. This demonstrates poor risk management practices on the part of the economic regulator's consultant and a very different orientation towards risk than that adopted by the regulated business. The expenditure proposed by the regulated business was largely disallowed.

Two other similar case studies illustrated very different interpretations between a consultant's understanding—accepted by the economic regulator—and a gas supply company's understanding of the requirements of the risk assessment frameworks of the relevant industry standards.

The case study of mains replacement programs for three gas distribution networks demonstrated that the economic regulator's decision to approve less capex than proposed by the businesses rested upon an interpretation of mandatory safety expenditure. The economic regulator concluded—based on its consultant's advice—that the volume of mains replacement specified in a safety case is indicative rather than mandatory (in fact, discretionary), and there are alternatives to mains replacement. This means the economic regulator considers that because there are options to manage mains risk, and the past volume replacement rates were well below forecasts, a company has discretion to manage mains risk.

Why does an economic regulator use external consultants? The answer lies in resourcing. Company interviewees were consistently critical—and often quite strongly—of the capacity and expertise of the economic regulator. Inexperienced staff, understaffing, high staff turnover, and poor operational and business understanding of analysts were constant themes in company interviewee comments who considered that these factors negatively impacted company processes and outcomes.

In summary, the economic regulator makes extensive use of external technical expertise because there are no equivalent internal resources. The availability of these specialised consultants is limited and many work for the gas supply companies which precludes them from providing expert advice to economic regulators due to the potential conflict of interest. For those consultants engaged by economic regulators, a different conflict of interest arises. These consultants are acutely aware of the economic regulator's position that a company's proposed expenditure should be reduced and produce advice to that effect. This situation is confirmed by our interview data and in-depth case studies of approved expenditure for regulated gas supply systems.

9.7 Discussion

Modern corporate regulation includes self-regulatory mechanisms that draw on private sector expertise. Peer pressure can provide an enforcement mechanism that may encourage and persuade industry actors to meet established norms and ensure that some companies that would otherwise have lower standards operate closer to what is typical [1, 2]. Despite this possibility, we see that use of consultants by economic regulators is not providing such a mechanism.

The two state-based regulatory mechanisms—economic and technical—are in conflict given their different objectives—lowest consumer price in one case and acceptable risk to the public in the other. This is problematic in its own right. What we have seen, however, is that use of consultants is exacerbating this division. As technical experts, consultants have the potential to contribute to resolution of the appropriate balance between these multiple objectives but they do not act in this way. Rather, we see them adopting one side of the conflict—that of the economic regulator—which aligns with their own financial interests. Such market-driven behaviour is to be expected, but it is not in the interests of the best regulatory outcomes overall. Compared to the handful of studies of similar arrangements [5, 9, 10], the benefit of increased economic regulatory expertise is not being seen but the downside of conflict of interest is certainly apparent.

Despite this, the economic regulator relies on the advice of external consultants over the expertise of the regulated businesses. Economic regulatory determinations over the last ten years or so have seen the views of consultants prevail over the views of the regulated businesses and even the technical regulator. In 50% of cases,[1] we found that the economic regulator approved significantly less or no safety-related capex proposed by the regulated business. This puts companies in the position of deciding whether or not to go ahead with safety-related expenditure.

In summary, the use of consultants by economic regulators has brought an imbalance between technical and economic goals as a result of the advice of consultants with insufficient expertise and/or a desire to support the economic regulator's position. This matters because, although neither the expert consultant nor the economic regulator have accountability for technical safety decisions, the economic regulator

---

[1] 50% of 24 economic regulatory determinations for gas supply businesses from 2010 to 2017.

unequivocally accepts the external consultants' advice and their views thereby can influence safety expenditure by gas industry duty holders.

9.8   Conclusion

Regulatory research has generally focused attention on the moves to risk-based regulation and forms of co-regulation. This includes some cases where regulatory functions are moved to the private sector exemplified by inspectors in the building construction sector. In this chapter, we have discussed a different category—where regulatory functions are retained by public agencies, but the actual delivery of those functions depends on the outsourcing of advice to private sector consultants.

Evidence from our research suggests that use of consultants by an economic regulator can create significant tensions with technical regulation compliance. In this case, the economic regulator's use of consultants is publicly available knowledge because the economic regulatory process produces large volumes of publicly available documentation including consultant reports. Use of consultants in other regulatory contexts is less visible, and so the impact on regulatory integrity is not as open to scrutiny.

Acknowledgements This work was funded by the Energy Pipelines Cooperative Research Centre, supported through the Australian Government's Cooperative Research Centres Program. The cash and in-kind support from the Australian Pipeline Industry Association Research and Standards Committee is gratefully acknowledged.

Ethical Statement This work was approved by the University of Sydney Human Research Ethics Committee (Project no. 2016/948). Informed consent was obtained from participants, and all data has been anonymised.

References

1. I. Ayres, J. Braithwaite, *Responsive Regulation: Transcending the Deregulation Debate* (Oxford University Press, New York, 1992)
2. R. Baldwin, J. Black, Really responsive regulation. The Modern Law Rev. **71**(1), 59–94 (2008)
3. R. Baldwin, M. Cave, M. Lodge, *Understanding Regulation: Theory, Strategy and Practice*, 2nd edn. (Oxford University Press, Oxford, 2012)
4. N. Gunningham, Safety regulation and mining inspectorates—lessons from Western Australia. J. Occup. Health Safety, Australia New Zealand **21**(4), 299–309 (2005)
5. J. Hackitt, *Building a Safer Future Independent Review of Building Regulations and Fire Safety: Final Report* (2018)
6. J. Hackitt, *Building a Safer Future Independent Review of Building Regulations and Fire Safety: Interim Report* (2017)
7. J. Hayes, L. Chester, D. Kramnaimuang King, ASME transactions, in *Proceedings of the 12th International Pipeline Conference (IPC 2018), Calgary, Alberta, Canada, 24–28 Sept. 2018*
8. A. Hopkins, Beyond compliance monitoring: new strategies for safety regulators. Law & Policy **29**(2), 210–225 (2007)

9. J. van der Heijden, On peanuts and monkeys: private sector involvement in Australian building control. Urban Policy Res. **28**(2), 195–210 (2010)
10. J. van der Heijden, Smart privatization: Lessons from private sector involvement in Australian and Canadian building regulatory enforcement regimes. J. Comp. Policy Anal. **12**(5), 509–525 (2010)

Chapter 10
Outsourcing in Theory and Practice: Insights from Nuclear Risk Governance

Jérémy Eydieux

Abstract This chapter examines two cases of risk governance in which actors interpret outsourcing as a possible source of operational vulnerability while using it to
strengthen safety governance. We propose to study this discrepancy through the
lens of organisational hypocrisy, suggesting a pragmatist approach as a means of
analysing hypocrisy in day-to-day managerial situations.

Keywords Outsourcing · Risk governance · Pragmatist approaches

10.1 Introduction

This chapter examines the "technical dialogue" used in France to govern nuclear
risk.[1] In order to secure authorisation for the decommissioning of a nuclear facility,
a nuclear operator (NO) has to produce a safety demonstration.[2] This document
is assessed by the regulator's technical support organisation (TSO), and then the
regulator decides how decommissioning should be carried out.

We study two cases related to two decommissioning projects in which (1) the
decision on whether to outsource or internalise part of the decommissioning operation
is seen as significant and (2) each organisation—the NO and the TSO—outsources
part of its contribution to the technical dialogue. The presence of outsourcing in both
cases leads to a discrepancy between the words and deeds of field actors. Both the

---

[1]This dialogue is a key process in nuclear risk governance in France, consisting of an exchange
of documents, information and ideas between experts from a nuclear operator and the technical
support organisation. Authorisation requests are one application of this process.

[2]In line with French regulations on the matter, this chapter uses "safety demonstration" to denote
an exhaustive collection of the safety guidelines scheduled by an NO for a project. It is often
communicated in digital format, as full demonstrations would often require a large cupboard if
stored in paper format. More generally, the word "safety demonstration" is often used to refer to
the entire process culminating in the regulator's decision.

J. Eydieux (✉)
Univ. Grenoble Alpes, Grenoble INP, CERAG, Grenoble, France
e-mail: jeremy.eydieux@grenoble-inp.fr

© The Author(s) 2022
J. Hayes and S. Tillement (eds.), *Contracting and Safety*,
SpringerBriefs in Safety Management,
https://doi.org/10.1007/978-3-030-89792-5_10

NO and the TSO interpret outsourcing as a possible source of vulnerability for the safety management of the decommissioning operations, while simultaneously using it to strengthen safety governance through the technical dialogue.[3]

We propose to study this discrepancy through the lens of organisational hypocrisy (OH), a process through which organisations satisfy their stakeholders with their discourse when they cannot do so with their actions. The literature on the subject does not clarify how OH works in day-to-day managerial situations, and we aim to fill this gap. Furthermore, prior research on OH has not yet looked at the issue of outsourcing, even though this is a controversial topic; we aim to focus on how OH works in situations involving outsourcing. Our intention in so doing is to help practitioners avoid the pitfall of OH through a better understanding of its mechanisms.

10.2   Organisational Hypocrisy: A Definition and Two Gaps

OH, which is often also referred to as "organised hypocrisy" [3], describes a process that enables organisations to survive when they are subjected to conflicting forces trying to affect their work. Hypocrisy allows organisations to satisfy all their stakeholders through their discourse by engaging in doublespeak, whereby their actions do not match their words. For example, an automobile manufacturer such as Volvo, "whose production seriously pollutes the environment", secures its own survival when it "proclaims the importance of working for a good environment and claims that its goal is to do so", rather than "preaching that pollution is necessary, that it is not dangerous, or even that it is something which we must accept" [4, p. 9].

OH is an interesting concept when considering the consequences of outsourcing. Indeed, outsourcing is likely to amplify any discrepancies between an organisation's discourse and its contractor's actions. OH becomes even more interesting when that discourse and those actions are related to industrial risk. For example, some construction companies are seen to promote their safety policies while failing to ensure safe working conditions for their contractors, or, even worse, while asking their contractors to downgrade their safety policies in order to reduce costs.

The literature on OH has not investigated the issue of outsourcing even though it is a controversial topic, as we can see in risk governance. In that field, practitioners have to determine the strengths and weaknesses of outsourcing specifically in relation to safety and risk. Since the 1970s, risk governance researchers have participated in lively academic debate around outsourcing. In addition, Brunsson [5, p. 124] observes that OH is "a way of handling situations when what is said cannot be done and when what is done cannot be talked about". However, the literature on OH does not clarify how this process can be observed in day-to-day managerial situations.

---

[3] We use the term "risk/safety management" to refer to the management of risks inherent to risky activities, and "risk/safety governance" to denote the management of risks inherent to risk/safety management.

In order to fill these two gaps, we aim to answer the following question: *How does OH translate into managerial situations, and to what extent does it explain the discrepancy between organisational discourse and actions in the context of risk management in cases of outsourcing?* Dualisms such as ideology and action, politics and action, or thought and action underlying Brunsson's reasoning around OH [7] are obstacles to the understanding of hypocrisy in managerial situations. We therefore propose an approach based on the pragmatist philosophical trend, which tends to move beyond dualisms.

10.3   Situated Organisational Hypocrisy and Outsourcing

For two decades, a wide array of disciplines has been reinterpreting the century-old philosophical trend of pragmatism in order to study real-life situations [16]. Pragmatism proposes an understanding of situated action that has been historically marginalised and that remains far from the mainstream in management science [13]. From a pragmatist standpoint, it makes little sense to rely on the dichotomy Brunsson draws between talk and actions [3–5, 7], as every statement made is a discursive activity and "action is a form of discourse" [13]. In order to simplify our reasoning, we therefore look for discrepancies between actions. As we study OH, we seek to identify discrepancies between actions connecting actors with an organisational practice—in our case, risk governance actors and the practice of outsourcing.

Dewey identifies a few ways by which actors may engage in an organisational practice—in our case, in outsourcing. First, they may rationalise how the outsourcing is supposed to work (beforehand or afterwards). Dewey identifies two types of ratio-nalisation [9]. On the one hand, actors may overemphasise the fact that outsourcing creates social asymmetries, which they then attempt to eliminate. This is what we call the "social approach". In this approach, outsourcing is experienced as creating and maintaining power relations between groups, as side effects of organisational processes or intentionally aimed at by actors; these asymmetries in turn create operational problems. On the other hand, actors may overemphasise the successes obtained through outsourcing. In this approach, which we call the "economic approach", actors engage in outsourcing as if it were solely driven by economic forces. Here, the purpose of outsourcing is to improve the whole organisation's economic performance through an adjustment to the distribution of operations management—in other words, through the make-or-buy decision.

According to Dewey, actors may also engage with outsourcing by making inquiries about it. Inquiries are sequential processes. First, actors experience an uneasy feeling about the outsourcing relationship, and transform it into a question such as "is outsourcing useful here?", in particular, "what do we learn from it?", "how can I better contribute to outsourcing?", or "why don't we involve this or that actor in the organising of outsourcing?" Then, they change their own contributions to the conversations and texts forming the outsourcing relationship, in order to fully interact with it. After reflecting on their experience, actors come to engage with outsourcing in a way that is more nuanced and "again … integral and vital" [9, p. 136]. We call this the "pragmatist approach", as, according to Dewey, it leads to the beliefs that are best aligned with actual situations [9]. In this approach, outsourcing is experienced as a means for organisations to access more verifiable facts through interorganisational communication [14].

The literature on organisation studies that draws on pragmatism describes some of the resources needed to make inquiries. Actors first need resources associated with themselves: mandates (or at least authorisations) from their organisation, genuine doubts they themselves have experienced, and time to finish their inquiries [12]. Actors also need resources associated with the richness or scarcity of their connection with the outsourcing relationship [17]. On the one hand, they need to access the field in which outsourcing occurs and to participate in the conversations involved in the organising of outsourcing. On the other hand, they need to access the texts used to make outsourcing happen (e.g. contracts, purchase orders, memos, and emails) and to participate in the drafting of new texts of this kind.

We thus define situated OH as a situation where in some circumstances actors have the resources to make inquiries about outsourcing, while in other circumstances, the same actors have limited means to do so. We will answer our research question by tracing each approach in the actors' behaviour related to outsourcing. We explain it through actors' wealth or lack of resources to lead inquiries, in other words the richness or scarcity of their connection with the outsourcing relationship.

10.4 Research Settings and Cases

In France, in order to secure authorisation for the building, operating, or decommissioning of a nuclear facility, every NO has to produce a safety demonstration. This document should prove that the NO has thought of all necessary measures to ensure nuclear safety, consistent with the latest scientific and technological developments and within economically viable conditions. The safety demonstration is submitted to the regulator, which submits it to its TSO for assessment. There is then a technical dialogue between the TSO and the NO consisting of document collection, interviews, and activity observations of operational activities. At the end of the dialogue, the TSO produces an evaluation report summarising the NO's design choices, makes any necessary observations, and suggests further measures to be taken by the NO to ensure safety. The report is sent to the regulator, which finally decides how building, operating, or decommissioning should proceed.

We study two cases involving decommissioning projects. The first mostly follows the internal processes of a NO and is related to a technical dialogue prior to the production of a safety demonstration. The other mainly tracks the internal processes of the TSO and is related to a technical dialogue prior to the production of a safety assessment report. In both cases, the decision on whether to outsource part of the decommissioning operation is a significant one. Also, each organisation—first the

NO and then the TSO—outsources part of the technical dialogue and the production of the document resulting from it.

In the case relating to the NO, we address the topic of "specialty outsourcing" (*sous-traitance de spécialité*). The decommissioning operations call for highly specialised workers—rope access technicians—and the NO decides to internalise them in order to ensure their compliance with nuclear safety standards. At the same time, however, the NO project managers are not concerned by the decision to outsource part of their design work to rope access technician providers. Meanwhile, in the case relating to the TSO, we look at "capacity outsourcing" (*sous-traitance de capacité*). Here, the NO outsources a significant part of the decommissioning operation. That is something of a problem for the TSO's human and organisational factors (HOF) department, as the NO delegates not only the operations but also part of its nuclear safety management. At the same time, the HOF department has no problem with the fact that it outsources part of its investigation and writing to a freelance HOF expert.

In both of these cases, field actors interpret outsourcing as a possible vulnerability for the decommissioning operations but at the same time they use it to strengthen the technical dialogue. Brunsson would analyse this as a discrepancy between discourse on risk management and action on risk governance. Conversely, we feel that, in the context of risk governance, actors have resources to make inquiries about the outsourcing relationships they can access directly, while their inquiries are limited when they experience the decommissioning operations' risk management.

10.5   Data Collection and Analysis

The data consists mainly of working documents collected from field actors' archives [2]. We collected emails, meeting minutes, proof documents, interview records, analysis texts, and slideshows, and complemented these with interviews. For the first case, we collected 47 documents (30 related to operational outsourcing and seven to risk governance outsourcing) and conducted eight interviews (lasting 12 h 45 min in total). For the second case, we collected 357 documents (240 related to operational outsourcing, 12 to risk governance outsourcing) and conducted three interviews (lasting 5 h 5 min in total).

Our study is based upon the technique of narrative analysis [8] using the framework proposed by Burke [6]. Instead of analysing the narratives through their common thread or structure, Burke proposes to analyse their ingredients: who (agent), what (act), how (agency), where and when (scene), and why (purpose). In this study, we analyse the use, as a method, of (a) the "pragmatist approach", where outsourcing is a means to share verifiable facts between several organisations; (b) the "social approach", where outsourcing is a means to create and maintain power relations between groups; and (c) the "economic approach", where outsourcing is a means to improve firms' economic performance.

10.6   Organisational Hypocrisy in Action

Our results reveal how OH works in managerial situations, where outsourcing is present in the context of risk governance. The findings are outlined in Table 10.1, which shows that, in both cases, actors face very similar situations, typical of our proposed definition of OH. When dealing with the outsourcing of part of their contribution to the technical dialogue, actors have resources to make inquiries, while these are limited when dealing with the decision on whether to outsource or internalise part of the decommissioning operation.

In the first case, where we look at the NO, designers realise in the early phases of the project that using 80 rope access technicians would be the best solution for disassembling the facility's equipment. However, the NO, although highly competent in decommissioning projects, has no experience in the personnel management of this type of worker. In order to assess the feasibility of the solution, the NO seeks advice from two companies managing rope access technicians. Their discussion reveals that the solution is feasible, but that maintaining the workers' skills can be a problem. Thus, the NO selects one of the two companies to take the discussion further, in order to improve its design of its disassembling operations and how these operations are explained in the safety demonstration. In these circumstances, the NO is in genuine doubt regarding its make-or-buy decision; it is able to access and participate in the conversations involved in the organising of outsourcing and to access and contribute to the texts used to make outsourcing happen. Thus, it makes inquiries about outsourcing and experiences them as a means to verify more facts. The NO practices the economic approach by going outside of its internal network— its industrial group—and the social approach by using power asymmetries to select one of the two initial contractors.

Table 10.1 Situated OH as a discrepancy caused by circumstances

		Engagement in outsourcing	
		Rationalisation: social or economic	Interaction: making inquiries
Circumstances of the relation to outsourcing	Activities underlying risk governance	Indirectly done	Done
	Future risky activities, as formalised in texts	Done	Not done

The NO's stance towards outsourcing is different regarding the rope access technicians in the safety demonstration. The NO chooses to internalise these in the hope of securing a stable, skilled workforce, because of the large number of workers involved and the mobility these workers usually exhibit within the nuclear sector. Outsourcing is seen as bad for safety in this situation. In its safety demonstration, the NO portrays the rope access technicians as the core of the disassembling operation and therefore critical for safe execution. While a lot of heavy-handling operations are involved, e.g. for slewing and depositing the heavy equipment, the HOF's analysis of disassembling focuses on rope access technicians. They are even the subject of a postulated worst-case accident scenario. In the circumstances of the demonstration (the text), the NO has limited opportunities to interact with the future rope access technicians. Thus, it focuses its management of outsourcing on its economic and social effects, internalising the technicians to ensure they are well covered by the HOF analysis. But at the same time, because it renders heavy-handling workers invisible, the safety demonstration portrays heavy-handling workers and rope access technicians as not interacting with each other.

Here, we have our first OH situation. In the way the design process is set up, NO actors have resources to make inquiries about the outsourcing of part of their design work and use them as a way to verify more facts. Their actions suggest that outsourcing can be positive. Yet, under the circumstances of the safety demonstration, actors are limited by the text-based medium. Thus, they play safe and internalise rope access technicians in order to limit what they say are the harmful structural effects of outsourcing.

In the second case, which focuses on the TSO, the NO anticipates that it will outsource a large part of its decommissioning operations. This is an important issue for the regulator, which asks its TSO to carry out an investigation into this issue; this is then delegated to the specialist HOF department within the TSO. The department's resources are limited and the topic is vast, so the expert associated with the investigation calls upon the services of a freelance HOF expert. The freelance expert helps design the questionnaire (to be sent to the NO), organise the fieldwork at the NO's site and collect documentation alongside this, write interviews and meeting minutes, and write a first draft of the analysis. Throughout the investigation, the TSO and the freelance expert interact as peers; for example, they deal with deliverables informally, considering a deliverable to be the document most recently produced by the freelance expert. In these circumstances, the TSO's HOF expert is able to access and participate in the conversations involved in the organising of outsourcing and to access and contribute to the texts that are used to make outsourcing happen. The TSO's expert also has incentives to organise the outsourcing from her organisation. This outsourcing helps the TSO to verify more facts, and the social and economic aspects of outsourcing are dealt with indirectly. To reduce asymmetries, HOF experts interact as peers, and to optimise the TSO's economic performance, the outsourcing is limited to the temporary need for another HOF expert.

As one might expect, the HOF experts, and thus the TSO, do not have the same position on the outsourcing anticipated by the NO for its decommissioning project. Thanks to the significant amount of evidence they collect, the HOF experts identify five harmful consequences of outsourcing. At the global level, they find it potentially dangerous that the site is simultaneously in production and in decommissioning and that it could become difficult in the future to maintain the skills needed. Regarding the effects of power asymmetries, they find that the HOF and ergonomic requirements are not discussed by the NO with its contractors and thus are not found in the contractors' documentation, and they find that supervision solely relies on checking documentation. The HOF experts focus their analysis on the harmful structural effects of outsourcing, thus contributing to risk governance with a helpful alternative viewpoint to that put forward by the NO. However, they are limited by the risk governance context, that is, by the questions asked by the regulator and within the TSO, and thus do not extend their criticism to the point of how the NO and its contractors will learn from each other, such as through meetings.

Here, we find our second OH situation. As part of the HOF investigation, the TSO's HOF department uses outsourcing positively to bring in external specialist expertise. In contrast, for the purposes of the safety assessment report, HOF experts are limited by the specifications initially outlined for their investigation and do not venture beyond the structurally harmful effects of outsourcing.

10.7   Normal Organisational Hypocrisy and Outsourcing

Throughout this chapter, we show how OH works in day-to-day managerial situations, especially within a context of risk governance and related to outsourcing. OH is a type of situation that connects actors to an organisational practice—in our case actors contributing to a specific type of technical dialogue and the practice of outsourcing. When the actors outsource part of their own contribution to the technical dialogue, they have resources to make inquiries about outsourcing, and use them as a way to verify more facts and indirectly manage the global structural effects of this practice. When the actors demonstrate or assess the safety of a decision related to outsourcing, their inquiries are limited and they only address the structurally harmful effects of outsourcing. In other words, in some circumstances, actors are able to experience the "pragmatist approach", while in others, they are limited to the "social" or the "economic" approaches.

We can therefore see how these actors specifically and involuntarily conform to the injunction to "do what I say, not what I do". However, as we define situated OH as a discrepancy between actions regardless of their discursive nature, we can account for discrepancies among discourses or among actions, while in Brunsson [3–5], OH is understood only as a discrepancy between talk and action. Conversely, we do agree with Brunsson's acceptance of OH, which is sometimes viewed as fatalistic [7, p. 295]. In our cases, actors do not *choose* to be hypocritical; they are *made* hypocritical by their circumstances, i.e. by the fact that they cannot be everywhere while outsourcing is everywhere. Hence, our understanding of OH cannot account for situations in which actors deliberately choose to lie and another pragmatist approach may be useful on this subject. This philosophical trend defends an instrumental understanding of truth, as shown by its utilisation in the study of the production, maintaining, and removal of ignorance [11].

We draw four conclusions specifically about outsourcing. First, our understanding of situated OH as arising from circumstances suggests that OH about outsourcing is a threat indicator for risk governance activities. If actors deal with potentially dangerous outsourcing through the "economic" or "social" approaches, the governance of risk is probably hindered by the ignorance of risk expertise [15] or by the actors' "pretence of knowledge" [10]. OH should be used as a magnifying glass to better govern risk, by directing attention to the potentially dangerous uses of outsourcing. Second, we observe through our cases that outsourcing is so controversial that, when actors have to write a text that is binding upon their institution, it can lead them to produce rationalisations that do not tally with their real-life experiences (such as declaring that outsourcing is a bad idea when in fact they use it successfully). This should invite practitioners of risk governance to carry out comparative analyses between what they write and what they do, which would be an interesting process to observe for research purposes. Third, we observe that outsourcing is contentious enough to prompt risk governance actors to produce or collect a vast quantity of documents about it. However, these documents do not compensate for the actors' lack of possible live interactions with practices that are yet to come, which pushes them to rationalise how outsourcing will work. This is an invitation for more informal communication within risk governance, involving experimental trials integrated within the process of the technical dialogue. Finally, these two episodes of technical dialogue show that the authorisation request neutralises the asymmetry relating to who has access to the reality in the field. When future activities only exist on paper, the NO, the regulator, and the TSO are all ignorant of "the field". Since they all have to "talk about places they have never been" [1], perhaps they can co-construct those places, thus benefiting from the expertise of each organisation.

Acknowledgements My best thanks go to the editors, especially S. Tillement who helped me a lot in clarifying my ideas and in making a sharper contribution to the debate about outsourcing from a fieldwork which was not on outsourcing per se. Thanks a lot to the field actors who granted me access to their activity and helped me in contextualising my observations.

Ethical Statement This work adhered to the research ethics that are stipulated in the "RESOH Chair convention" that complies with relevant legislation regarding ethical conduct of research. Informed consent was obtained from participants, and all data has been anonymised.

References

1. P. Bayard, *How to Talk About Places You've Never Been* (Bloomsbury Libri, London, 2016)
2. G.A. Bowen, Document analysis as a qualitative research method. Qual. Res. J. **9**(2), 27–40 (2009)
3. N. Brunsson, *The Organization of Hypocrisy: Talk, Decisions and Actions in Organizations* (John Wiley & Sons, Chichester, 1989)
4. N. Brunsson, The necessary hypocrisy. Int. Exec. **35**(1), 1–9 (1993)

5. N. Brunsson, *The Consequences of Decision-Making* (Oxford University Press, New York, 2007)
6. K. Burke, *A Grammar of Motives* (University of California Press, Berkeley, 1945)
7. P. Carter, Nils Brunsson: the organization of hypocrisy: talk, decisions and actions in organizations. Organ. Stud. **13**(2), 291–295 (1992)
8. I. De Loo, S. Cooper, M. Manochin, Enhancing the transparency of accounting research: the case of narrative analysis. Qual. Res. Account. Manag. **12**(1), 34–54 (2015)
9. J. Dewey, *Individualism, Old and New* (Unwin Brothers, London, 1931)
10. S. Ghoshal, Bad management theories are destroying good management practices. Acad. Manag. Learn. Educ. **4**(1), 75–91 (2005)
11. M. Girel, *Science et territoires de l'ignorance* (Éditions Quae, Versailles, 2017)
12. B. Journé, N. Raulet-Croset, La décision comme activité managériale située. Une approche pragmatiste **225**, 109–128 (2012)
13. P. Lorino, *Pragmatism and Organization Studies* (Oxford University Press, Oxford, 2018)
14. P. Lorino, D. Mourey, The experience of time in the inter-organizing inquiry: a present thickened by dialog and situations. Scand. J. Manag. **29**(1), 48–62 (2013)
15. H. Merkelsen, Institutionalized ignorance as a precondition for rational risk expertise. Risk Anal. **31**(7), 1083–1094 (2011)
16. B. Simpson, F. Den Hond, The contemporary resonances of classical pragmatism for studying organization and organizing. Org. Stud. **2021**, 1–20 (2021)
17. J.R. Taylor, E.J. Van Every, *The Emergent Organization, Communication as Its Site and Surface* (Lawrence Erlbaum Associates, Mahwah, 2000)

Chapter 11
Outsourced Enforcement: Improving the Public Accountability of Building Inspectors

Nader Naderpajouh, Rita Peihua Zhang, and Jan Hayes

Abstract Regulatory enforcement of building safety and quality has been outsourced with a move to partially privatised building inspectors in both the UK and Australia. The Grenfell Tower fire and other near misses in Australia highlight the problems this has introduced. This chapter reviews the role of building inspectors using a public administration accountability framework and recommends structural changes to improve safety for occupants of high-rise buildings.

Keywords Public accountability · Building inspection · Building regulation · Privatisation · Grenfell Tower

11.1 Introduction

The high-rise building construction sector has seen several high-profile accidents in recent years in the UK, Dubai, China, France, South Korea and Australia. The consequences of these events (both realised and potential) demonstrate that high-rise buildings are effectively a kind of hazardous infrastructure and yet regulation of safety in high-rise construction has more in common with domestic building construction than with regulation of safety in other hazardous industries. Along with trends to performance-based regulation, a recent key factor has been privatisation of building inspection activities where inspection has been effectively outsourced from government to the private sector. Multiple accident investigations and industry reviews have noted this as a problem and yet solutions remain elusive.

This chapter draws on theories of public sector accountability to show that the issues are structural and so not amenable to solutions simply targeting skills and competency of building inspectors individually as has typically been the case.

N. Naderpajouh (✉) · R. P. Zhang · J. Hayes
RMIT University, Melbourne, Australia
e-mail: nader.naderpajouh@sydney.edu.au

N. Naderpajouh
The University of Sydney, Sydney, Australia

© The Author(s) 2022
J. Hayes and S. Tillement (eds.), *Contracting and Safety*,
SpringerBriefs in Safety Management,
https://doi.org/10.1007/978-3-030-89792-5_11

11.2 Effective Regulation and Effective Regulators

The building sector is controlled by regulations which are legal instruments intended to ensure provision of socially acceptable levels of health, safety and welfare of building occupants and users [7]. As such, building regulations specify regulatory controls over the design, construction, operation and demolition of buildings, addressing various aspects such as structural integrity, fire safety, heating, lighting, ventilation, sanitary facilities and indoor air quality [7]. Traditionally, building regulations were prescriptive and expressed as detailed descriptions of how each building component or system must be constructed.

Many countries have transitioned from a prescriptive-based building regulatory regime to a performance-based building regulatory regime [13]. A performance-based building code specifies the criteria to be achieved as an outcome, rather than prescribing the specific manner in which a building is to be constructed. Parallel to the common justifications of efficiency in neoliberalism, the underlying motivation of such changes in the building industry included "reducing regulatory burden, reducing costs to the industry and public, increasing innovation and flexibility in design and construction, and being better positioned to address emerging issues" [6].

Regulated standards for construction are one thing, but in a competitive environment, regulation is of limited effectiveness without some degree of common oversight and enforcement activities to provide construction companies with an incentive to comply. As has been found in other industries, moving to performance-based regulation requires highly skilled and motivated regulators, given the changed role of regulatory enforcers from inspecting specific items to certifying performance.

The move to performance-based regulations has often come about in parallel with changes to the system for appointing building inspectors who are responsible for certifying firstly that the building design meets code requirements and secondly that the building is safe to be occupied once it is constructed.[1] What was once a function of government employees, building inspection has been privatised as a profession despite inspectors themselves retaining public accountability for building quality. This change has occurred in multiple jurisdictions including Australia and the UK as we will describe further below.

11.3 Public Accountability

Accountability is fundamental in public administration. Public servants are answerable for their performance, but questions arise as to which stakeholders are legitimate sources of control and what are legitimate performance expectations. Traditionally,

---

[1] In some jurisdictions, these tasks are performed by a building surveyor (rather than inspector). Interim physical inspections during the construction process may be performed by a contracted building inspector working for the surveyor. We use the term "inspector" to mean the person with the legislated responsibility for checking compliance and issuing permits.

public servants were answerable to parliament and the courts, but this is no longer the case. It has been pointed out that "in a complex administrative state, characterised by widespread delegation of discretion to actors located far from the centre of government, the conception of centralised responsibility upon which traditional accountability mechanisms are based is often fictional" [10, p. 38].

To address this complexity, Romzek and Ingraham [9] propose a four-factor typology of public accountability that supports investigation of regulatory action through an accountability lens. The public administration accountability framework comprises legal, bureaucratic, professional and political accountability. Building inspectors are no longer public servants; i.e., they are not employed by government, and yet they fulfil a statutory function, enforcing the building regulations by certifying firstly, that the design and later, the building itself, meet the relevant codes. As such, the public interest is fundamental to their activities and the public administration accountability framework provides a useful perspective on how their role is defined and how it works in practice. Note that in the context of building inspectors, the framework is being applied to study the inspectors' accountability for their actions, not the compliance of industry with regulations which is a separate issue.

Legal accountability describes how the work of the building inspectors is driven by statutory rules defining their role, responsibilities and mandated qualifications. Bureaucratic accountability is concerned with other sets of rules that are used by the building inspectors in carrying out their work. These are the regulations and codes that the building sector must comply with. Building inspectors draw on these in assessing compliance and so issuing permits. Professional accountability drives building inspectors to exercise their professional judgement in the regulation implementation process in accordance with professional norms. Examples of mechanisms to ensure professional accountability include codes of conduct.

Finally, political accountability is the responsiveness of building inspectors to political constituencies which may include the general public, building occupiers, the construction sector generally, elected officials, statutory agencies or other arms of government, future generations and of course their client who is funding their inspection services. Political accountability aims to ensure participation and approval of those who are regulated and society at large as a form of feedback to the regulatory system [1]. As Romzek and Dubnick point out, "while political accountability systems might seem to promote favoritism and even corruption … they also serve as the basis for a more open and representative government" [8, p. 229]. In the context of building inspectors, their private sector status brings a strong implication that political accountability to their clients is significant as we shall see.

The process of assessing accountability performance requires three phases: (1) provision or obtaining information (the accountable party reporting on their actions), (2) review and assessment (discussion and/or evaluation of actions taken) and (3) consequences (reward or sanction based on performance) [1]. Whilst the forum that deals with these phases may vary according to specific circumstances, all three are required and a culture of accountability may fail in the absence of established information provision processes, credible and ratified review process, as well as lack of consequences if accountabilities are not executed appropriately.

11.4 Grenfell Tower

On 14 June 2017, a fire broke out in the 24-storey Grenfell Tower building in West London, resulting in 72 deaths and more than 70 injuries. Investigations revealed that the fire started with a malfunctioning refrigerator on the fourth floor, but it spread quickly to the entire building as a result of recently retrofitted combustible cladding on the building exterior [3].

In the UK, compliance with the relevant building regulations is primarily the duty of the person carrying out the building work, but, unless the work is defined as low risk, independent building control oversight is required. This was the case for the Grenfell Tower modifications to install new cladding. This service can be provided in two ways. The duty holder can either approach their local authority (i.e. local government) and have them provide the building control service for a fee or choose the approved inspector route. In the second case, a private sector practitioner is hired to provide inspection services on a commercial basis. The inquiry into the UK construction regulation following the Grenfell Tower disaster highlighted that "part privatisation of this regulatory function has created a unique competitive environment and has introduced unintended consequences" [4, p. 54]. Most authorised inspectors work for small organisations that rely on their relationships with builders and building construction companies for their ongoing work. The accountability framework discussed above provides a way to examine these arrangements.

Legal accountability is weak with no oversight of the quality of inspectors' work in place and "no legislative requirements that set standards of competence or training for building control inspectors" [4, p. 55], either in the public or private spheres. Bureaucratic accountability is facilitated by the large volume of codes and standards and the drawings submitted to inspectors for review, so in most cases there will be large volumes of data available for them to consider. However, the lack of any checking or audit mechanism for inspection activities means that there are typically no consequences for building inspectors if they fail to correctly consider this information. Professional accountability is also weak but increasing, with efforts afoot to professionalise these workers, including development of professional standards and formal tertiary qualifications. Professional standards would not only clarify for building inspectors exactly what is required; they would provide the basis for auditing building inspection activity and so enable consequences of accountability failings to be instituted.

Currently, building inspectors are largely driven by political accountability, i.e. direct consideration of stakeholder interests. This is not necessarily problematic, depending on whose interests prevail. The inquiry into UK building regulation post-Grenfell [4] found that, while introduction of this arrangement had driven down costs, inspectors have a conflict of interest and experience a "difficult trade-off", failing to gain work if they do not approve more risky designs. Inspectors no longer act as impartial checkers of building safety, but rather "become far too embedded in supporting the building design process rather than being an impartial rigorous verifier of buildings safety" [4, p. 54]. The inquiry also reported "repeated concerns expressed about the commercial pressures associated with rigorous enforcement of fire safety requirements" [4, p. 55]. In summary, in many cases, the inspectors are effectively accountable to the very parties whose work they should be checking for compliance and face significant consequences if they fail to act in accordance with this accountability. This is not consistent with the best public safety outcomes and is an indicator of a systemic accountability gap in the sector.

Outsourcing of building control and other systemic inspection problems contributed significantly to the loss of 72 lives at Grenfell. The modifications to Grenfell Tower that covered the refurbished building in flammable cladding were approved by a building inspector working for the local authority. In evidence to the ongoing formal inquiry into the circumstances of the disaster, the inspector has admitted that his work fell below the standards of a reasonably competent inspector,[2] yet he pointed out that at the time he was responsible for reviewing up to 130 projects.[3] Site inspections were carried out on behalf of the local authority by a contracted inspector who said his remit was not to check whether work met regulations[4] or matched architects' drawings but rather to check that the work was neat and tidy.[5]

At the time of writing, the Grenfell Inquiry is ongoing. Privatisation of building inspection is only one of the weaknesses in the existing UK building quality control framework that has been revealed by the disaster. One key structural improvement is the imminent move of regulatory responsibility for buildings to the UK Health and Safety Executive which successfully performs this function for many other hazardous industries. It can only be hoped that this heralds a significant cultural and structural shift in public accountability for building quality in the UK.

11.5 Incidents in the Australian Building Sector

There have been two significant fire events in Australian high-risk buildings in recent years. No fatalities or injuries have resulted, but these are all near miss incidents that could have had much more serious consequences similar to the Grenfell case. The Lacrosse Building fire in Melbourne in 2014 affected hundreds of occupants who needed immediate evacuation and accommodation [2]. In early 2019, fire spread along the side of the Neo200 building (also in Melbourne) which used combustible cladding similar to the material used in the Grenfell Tower. Problems are not confined to Melbourne or to cladding. In 2018, hundreds of residences of the 36-storey Opal

---

[2] Transcript of Proceedings, Grenfell Tower Inquiry (Day 46, Sir Martin Moore-Bick, October 1 2020), 100 (J. E. Hoban).

[3] Transcript of Proceedings, Grenfell Tower Inquiry (Day 45, Sir Martin Moore-Bick, September 30 2020), 111 (J. E. Hoban).

[4] Transcript of Proceedings, Grenfell Tower Inquiry (Day 42, Sir Martin Moore-Bick, September 24 2020), 153 (J. White).

[5] Transcript of Proceedings, Grenfell Tower Inquiry (Day 42, Sir Martin Moore-Bick, September 24 2020), 171–172 (J. White).

Tower apartment building in Sydney were evacuated on Christmas Eve because structural cracking was observed in the building [12].

Building regulation is a state responsibility in Australia so there are variations in the details between states, but the states where the above events occurred operate systems that are very similar to that in the UK described above. Regulations are performance-based, and building inspection has been partly privatised. Building inspection can take two routes at the discretion of the builder, and certification is required, covering both the design and the completed building. Furthermore, increasing privatisation of regulatory enforcement has been found to result in a decline of accountability [13]. Following the Lacrosse fire and in the knowledge of other problems regarding counterfeit materials in the building supply chain, a national review of compliance and enforcement systems in building and construction was instigated in 2017 [11]. The review found multiple issues with the building inspection system that can be usefully summarised using a public accountability framework.

Shergold and Weir highlighted major problems with the legal accountability of private building inspection activities saying, "it is not just the conduct of private building surveyors that contributes to the problems, but also the lack of regulatory oversight of their conduct and, more importantly, the absence of a cohesive and collaborative relationship between state and local government and private building surveyors" [11, p. 14]. When it comes to bureaucratic accountability, they reported that decisions are based on documentation provided but, in practice, preparation of as-built plans is far from the norm and, as such, occupancy permits are often granted on the basis of out of date information. This means that changes to the approved design occur frequently without independent certification.

Professional accountability has also been weak. The review noted that "[m]any private certifiers are individuals of high integrity" [11, p. 14]. But they have acted as individuals with little support provided by the system. They face political accountability in addressing "the conflicting demands they face from their clients, the regulators and the insurers" [11, p. 14]. The review was told that when inspectors try to undertake enforcement action they are not supported by government and "on occasion, attention turns to their own conduct and they find themselves the subject of complaint and criticism" [11, p. 14].

With the strongest accountability being to their clients who they depend on financially, it is hardly surprising that investigations into specific incidents reveal a lack of regulatory enforcement. Following the Lacrosse apartment fire mentioned earlier, the building surveyor was referred for disciplinary action to the Building Practitioners Board on the basis that he could not have been satisfied that the building work met the relevant regulatory requirements when he issued the building permit. After various legal proceedings, he was found to be liable for the cost of repairs to the building, along with the architect and the fire engineer.

To address these issues, the review recommended statutory controls to mitigate conflict of interest, development of a code of conduct for building inspectors, mandatory reporting obligations regarding regulatory breaches and increased collaboration between building agencies and privatised inspectors on enforcement matters.

Changes are being made in these areas in various states including new bureaucratic accountability support by issuing templates for inspectors to use, a new code of conduct providing not only professional norms but also a basis for enforcement, i.e. consequences for accountability failure. It is too soon to be able to assess the impact of these changes, but viewed through the framework of accountability types and processes, they are moves in the right direction.

11.6 Discussion

As we know from studies of regulation in other sectors, having skilled and competent regulators is critical in successfully enforcing performance-based regulation. In the building sector, we find that despite its criticality in ensuring public safety in high-rise apartments, building inspection is not a popular career choice. In Australia and the UK, the average age of building inspectors is over 50 and there is no adequate career pathway leading to this important role. Given the current arrangements, it is easy to see why this is not a popular career choice within the construction sector. This needs to change to attract the best people to this critical role.

Too often in the wake of accidents, the focus has been on building inspectors themselves as we see with the liability findings in the Lacrosse case. Our analysis suggests that the problem is more structural. If building inspectors are to successfully take on the challenge of their important role, they must be better supported. There is a need to create a simple structure of public accountability for building inspectors that draws on all four public accountability types.

From a legal accountability perspective, the place of the building inspectors needs to be reinforced in legislation, including their authority to take appropriate enforcement action. Bureaucratically, there must be procedures in place for government agencies to support their frontline enforcement representatives, i.e. building inspectors, in doing their job. If local government is not able to provide such support, then there is a case for larger structural changes as we have seen in the UK with building regulation moved to an agency with more experience in regulating system safety in hazardous industries. In addition to support, comes the need for formal auditing of performance in some way to ensure that enforcement standards are maintained.

When it comes to professional accountability, improvements are also needed including articulation of a clear set of professional standards and expectations and a system of peer support so that building inspectors are encouraged to act in a way consistent with the standards set. Specific training and licensing arrangements linked to professional standards would allow the costs associated with building inspection work to be linked to specific requirements and defended in the case of commercial pressure.

Since privatisation, building inspectors' primary political accountability has been to commercially focused building sector stakeholders. The changes suggested above are designed to rebalance the system. The safety of high-rise building occupants must play a stronger role in driving day-to-day activities of building inspectors. This will only happen if the work of building inspectors is overseen more closely by government, so the necessary political accountability is in place.

These problems have arisen in parallel with the part privatisation of building inspection, but if sufficient support and oversight is provided to ensure that enforcement standards are being maintained, there is no inherent reason why these inspection activities cannot be outsourced to the private sector. When it comes to enforcement, low-level activities such as persuasion and issuing notices can be done by licenced private sector professionals but strong links are needed into government to deal with cases that require escalating enforcement action and to ensure standards of inspection are met.

The justification for such changes is not only safety-related. When we look at building quality more broadly, the building sector as a whole is riddled with problems. A recent study of 212 residential multi-occupant building audits in Australia suggested a staggering 85.7% of the buildings had at least one identified defect with an average of more than 14 identified defects per building [5]. Reform is needed across the sector, but a strong enforcement system is a key part of any overall plan to lift building quality for both safety and commercial reasons.

11.7 Conclusion

The Grenfell Tower disaster must be a turning point in raising awareness about the problem of building safety. This case highlights many weaknesses in the sector, but here we have focused on outsourcing of the regulatory function of building inspection using the theoretical lens of public accountability. In effect, the building sector in both the UK and Australia has operated with little regulatory oversight and little enforcement since building inspection activities were partly privatised several decades ago. The reason lies not so much in the skills and competencies of individual building inspectors, but in the structure of the system that drives them to consider political accountability to other building industry stakeholders above all else as a result of weaknesses in other mechanisms for ensuring public accountability. This occurs despite clear evidence (in the form of accidents and near misses) that this approach is not adequate, particularly in relation to high-rise buildings. Building inspectors are left to work within a system in which weak public accountability mechanisms are inherent in the structure of the system itself.

As a result, there is a need to provide structural changes that ensure public accountability, with a focus on the end-users. This includes addressing factors linked to all four facets of public accountability—legal, bureaucratic, professional and political. Providing the necessary support and oversight mechanisms, via legal, bureaucratic and professional means, can serve to rebalance the political accountability of building inspectors away from commercial building sector interests and further towards the health and well-being of building occupants and users. While outsourcing has significantly contributed to the current situation, if government processes can more strongly support and supervise private building inspectors, this aspect of the structure does not have to be inherently problematic.

References

1. G.J. Brandsma, T. Schillemans, The accountability cube: measuring accountability. J. Public Adm. Res. Theory **23**, 953–975 (2012)
2. G. Genco, *Lacrosse Building Fire 673 La Trobe Street, Docklands on 25 November 2014* (2015)
3. GTI, *Report of the Public Inquiry into the Fire at Grenfell Tower on 14 June 2017* (2019)
4. J. Hackitt, *Building a Safer Future Independent Review of Building Regulations and Fire Safety: Interim Report* (2017)
5. N. Johnston, S. Reid, *An Examination of Building Defects in Residential Multi-owned Properties* (2019)
6. B. Meacham, *Toward Next Generation Performance-Based Building Regulatory Systems* (2016)
7. B. Meacham, R. Bowen, J. Traw, A. Moore, Performance-based building regulation: current situation and future needs. Build. Res. Inf. **33**(2), 91–106 (2005)
8. B.S. Romzek, M.J. Dubnick, Accountability in the public sector: lessons from the challenger tragedy. Public Adm. Rev. **47**(3), 227–238 (1987)
9. B.S. Romzek, P.W. Ingraham, Cross pressures of accountability: initiative, command, and failure in the Ron Brown plane crash. Public Adm. Rev. **60**(3), 240–253 (2000)
10. C. Scott, Accountability in the regulatory state. J. Law Soc. **27**(1), 38–60 (2000)
11. P. Shergold, B. Weir, *Building Confidence: Improving the Effectiveness of Compliance and Enforcement Systems for the Building and Construction Industry Across Australia* (2018)
12. Unisearch, *Opal Tower Investigation Final Report* (2019)
13. J. van der Heijden, On peanuts and monkeys: private sector involvement in Australian building control. Urban Policy Res. **28**(2), 195–210 (2010)

Chapter 12
Implications for Safe Outsourcing

Stéphanie Tillement and Jan Hayes

Abstract This chapter describes some of the lessons highlighted by the different authors of the book. By assessing the different contributions, it discusses the main research and managerial challenges related to the nexus between safety and outsourcing practices. Two main issues are considered: (1) the implications of fragmentation and (2) the importance of addressing the temporal dimension and transience.

Keywords Outsourcing · Safety · Fragmentation · Temporalities · Transience

12.1 Introduction

This book was born out of the observation that in the face of an increased use of outsourcing in high-hazard organisations in a more and more varied range of activities, the effects of contracting on system safety were still little addressed. More importantly, looking at the debate and theory development in safety studies, we observed that the relatively few studies addressing this link were quantitative, rather static and distant from work situations and emerging and evolving collective practices. This is surprising given the recent developments of practice-based approaches in the field of safety [4, 5]. As such, we perceived that there is a need for research that is more empirically grounded, reflecting on and through practices, beneficial to both academics and practitioners and resulting from close collaborations between these actors.

All the chapters presented in this book rely on an in-depth understanding of the empirical field they are studying, while investigating very diverse outsourcing situations and thus each tackling specific issues. The chapters describe ten case

S. Tillement
IMT Atlantique, LEMNA, Nantes, France
e-mail: stephanie.tillement@imt-atlantique.fr

J. Hayes (✉)
RMIT University, Melbourne, Australia
e-mail: jan.hayes2@rmit.edu.au

© The Author(s) 2022
J. Hayes and S. Tillement (eds.), *Contracting and Safety*,
SpringerBriefs in Safety Management,
https://doi.org/10.1007/978-3-030-89792-5_12

studies regarding use of contracting in sectors from nuclear waste management to railways and the building sector and in lifecycle stages from design and construction to decommissioning and even regulation.

All cases suggest that the structure imposed by classical contracting (including separate organisations, fixed responsibilities and well-defined tasks) creates problems for those whose professional practice occurs in this environment. Operating a very hard line based on predefined definitions about who does what and who is responsible for what creates safety problems. This volume, focused on situations where part of the work is contracted out, uncovers problems that concern process and workplace safety more globally. Such issues deserve attention both from researchers and from practitioners. These problems arise in two interrelated areas: (1) the fragmentation of work and introduction of multiple boundaries and (2) the temporary nature of contracting arrangements and the significant dynamic, temporal dimension that is introduced. Given the relative lack of safety theory on which to draw, chapter authors have turned to various theories from other domains such as Science and Technology Studies (STS) and organisation studies to frame their work to bring answers to these problems and contribute to the literature. Examples include Russel and Tillement's use the concept of boundary objects [10], Naderpajouh et al.'s use of a public accountability framework [9], Eydieux's consideration of organisational hypocrisy [3] and Griegel and Gould's use of theory of temporary organising [6]. Some of these can illuminate aspects of the fragmentation issue, while others highlight the transient nature of outsourcing arrangements.

12.2   The Problem of Fragmentation

In this book, safety is conceived as the product of actions and interactive dynamics between various stakeholders, including clients and (sub)contractors. These dynamics are affected by the structures and socio-material entities in place. The fragmentation of work in the contracting environment challenges many of the organisational preconditions for excellent safety performance. Fragmentation narrows the focus of those down the contracting chain in ways that may mean that safety outcomes no longer receive sufficient attention. The cases regarding use of outsourcing in regulation show this. Hayes, Chester and King demonstrate that the narrow interests of consultants can jeopardise a focus on long-term safety outcomes in the field of energy regulation. Similarly, Naderpajouh, Zhang and Hayes show that outsourcing the role of building inspectors has contributed to deterioration in building quality.

In many of the case studies included in this volume, we see that informal practices have grown up to compensate for the boundary between organisations to try to meet work requirements in practice. In some cases, these are not good for safety. Dechy and Largier's case shows work being done in some cases by people without the necessary qualifications in response to late changes in resourcing requirements. At the other extreme, Hara's chapter describes the case of bullet train construction in which case boundaries between contractors have essentially been collapsed.

The accident case reviewed by Hara suggests that ignoring interfaces might have downsides too if responsibilities become opaque. The best response for safety would appear to be moving towards a sensible mid-point as is proposed by Gotcheva et al. in the context of nuclear power station construction. They propose use of alliance contracting where the contract provides a framework for cooperation, rather than a way of transferring risk to others. These new forms of contracting favours co-creation and joint cooperation.

But outsourcing does not only involve inter-organisational boundaries: the use of (sub)contractors, whether for permanent activities (e.g. maintenance) or for temporary projects, always requires important reorganisation by the principal which may affect internal boundaries. Russel and Tillement show that a lack of understanding across intra-organisational boundaries can make effective temporary organising even harder. Interfaces defined by people who do not understand the work activity, often because they are distant from it, can be a major problem. We see this in the outsourcing of nuclear power plant operations or in managing turn arounds. It may also affect planning and the maintenance of well-informed, adapted knowledge and skills (both for in-house and for external personnel).

In the face of fragmentation and the resultant multiplication of boundaries, the chapters highlight practices observed in the field that go some way to compensating for the negative effects on safety. This includes restructuring activity to provide more integrated work scopes, appointing specific boundary-spanning agents, e.g. "utility coordinator" (olde Scholtenhuis) or project leader (Russel and Tillement). Other potential solutions are mentioned, which should be the subject of further studies, such as aligning payment structures with desired behaviours (not just minimising time spent), establishing contract styles to combat the side-effects of price competition or strong institutional devices to support inter-organisational coordination.

Many authors also insist on the importance of building and sustaining long-term relationships between clients and contractors as a possible answer to the issue of fragmentation. This leads us to our second main point, the importance of transience and temporalities when it comes to understand and managing the nexus between outsourcing and safe industrial performance. As implicitly described in this section, the temporal dimension is key in understanding and managing the consequences of fragmentation such as shorter-term focus, mismatched or conflicting timing norms between stakeholders (e.g. contractor and principal or several contractors) or a focus on efficiency rather than long-term outcomes such as safety, sustainability or learning. This is discussed in more detail in the next section.

12.3 Transience and Temporality

Time has become a subject of scrutiny in the field of organisation studies [2], notably in so-called process studies, but is still largely ignored in the safety literature. Yet, as emphasised in many chapters in this book, the issue of time is crucial when it comes to understanding safe industrial performance. This is even more true when outsourcing becomes central in the way of organising work. This volume, through the empirical studies on which it is based, encourages reflection on the temporal dimension of outsourcing. Outsourcing is studied as a dynamic process, which involves actors and relationships that evolve over time. Either explicitly (e.g. Tillement and Leuridan; Helledal and Pettersen) or implicitly (e.g. Hara; olde Scholtenhuis), all the chapters focus on temporally evolving phenomena, whether they be the emergence and normalisation of outsourcing practices (Tillement and Leuridan), the temporary nature (or not) of relationships in project-based organisations (Russel and Tillement), temporariness or transience of work structures (olde Scholtenhuis; Helledal and Pettersen), the long-term or short-term temporal orientation of actors and its effect on the quality of interactions and trust relationships (Hayes et al.; Dechy and Largier).

The chapters highlight several temporal orientations or foci associated with outsourcing that tend to hamper safety. Many organisations tend to focus on clock-time emphasising schedules and delays, resulting in increased temporal constraints and production pressures all long the supply chain (olde Scholtenhuis; Dechy and Largier). Many contractors consider that the client is mostly attentive to direct and short-term goals such as efficiency and pays less heed to longer-term imperatives such as safety. Contractors describe having little support when it comes to ensuring such long-term outcomes and, in some cases, even see consideration of such long-term goals as beyond their remit. This finding is not specific to situations where part of the work is outsourced but appears exacerbated in that case. It is observed in classic activities such as maintenance or construction but also, more surprisingly, in regulation and governance activities (Hayes et al.; Naderpajouh et al.). This last element is made worse by transience and the temporary nature of relationships between actors, which complicates the establishment of informal arrangements negotiated to deal with immediate contingencies. This is despite such arrangements being necessary for effective learning and the establishment of clear jurisdictions. Paradoxically, the temporary forms of organising that often accompany outsourcing tend to reinforce bureaucracy with its well-known effects on strict reliance on rules and rigidity.

The chapters sketch some practices that may help in resolving the temporal tensions with which organisations are confronted. First, in the face of transience, establishing long-term relationships between contractors and clients reinforces trust along the supply chain, supports the building of shared and negotiated practices to face unexpected events, and sustains the transmission and maintenance of skills. Contracts have also a key role to play and are clearly related to the first point. The chapters in this volume advocate for less rigid forms of contracting, such as alliance contracting (Gotcheva et al.). The main idea is to build contracts that are negotiated locally and so are more able to take into account the specificities of the activities that are contracted out, notably the level of complexity (olde Scholtenhuis). Such arrangements can include performance indicators, penalties and incentives that go beyond economic performance and short-term results by including long-term goals. Finally, all that precedes also shows the importance of setting institutions or institutional arrangements, such as regular review meetings to make sure that these "ways of doing" remain robust and can be discussed and renegotiated to adapt to any contingencies of the "reality of the field" (Tillement and Leuridan, Hara).

12.4   Taking Stock and Looking Forward: Advancing Safety Theory

Since the late 1990s, an organisational view of accidents has become common thanks to the seminal research from James Reason regarding the Swiss cheese model and also the work of the high reliability researchers. Despite the focus moving from individuals to organisations, few researchers have acknowledged that, as a result of changing work practices, Rasmussen's nested levels of decision-making no longer apply [8]. There is no longer a neat, linear progression in the social organisation of safety from individuals and staff, to management, company, regulators and government. New approaches are needed to deal with the higher level of system complexity.

The cases taken collectively progress the debate about the link between structure and culture. Hopkins argues that structure creates culture [5]. His thesis is that decentralisation in the way organisations are structured allows cost to take precedence over safety. He contends that centralised structures where safety information is valued and can readily travel up to the levels at which key decisions are made lead to the most informed senior managers and the best safety decisions.

Hopkins' work is grounded in the study of large corporations but is silent on the subject of contracting. Contracting could be seen as an extreme form of decentralisation and therefore subject to all the associated problems for safety that he highlights. In the case of contracting, the financial performance of the entire chain is still linked and yet each organisation manages their work, including safety implications, independently (at least in theory). In this way, Hopkins' theory predicts that outsourcing will create safety problems and indeed this is the case. Looking for the practical implications, improved communication of safety issues across organisational boundaries and aligning the interests of all parties are broadly consistent with his recommendations on effective structures for individual organisations.

All chapters in this volume offer situated visions and results of the outsourcing–safety nexus, which make it impossible to qualify it simply as being always positive and negative. As discussed, this conclusion depends on many factors linked to the characteristics of the organisations and of work activities. Taken together, the collection of chapters identifies several key organisational features that are useful in assessing the forces and vulnerabilities of a network of organisations in terms of safety and organisational reliability. The conclusions proposed in this book are in line with previous work that has paved the way for the identification of key criteria for defining "safety regimes" [1]. They complement them by adapting them to more distributed and temporary settings in which the issue of outsourcing is central. Three major characteristics can be distinguished, which overcome the specificities of each industrial sector or activity, and which are tightly connected to the two main issues discussed in previous sections, i.e. fragmentation and also transience and temporalities.

The first issue concerns the formalisation and control of practices, jurisdictions and relations through contracts, rules, procedures and standards. Outsourcing leads to more decentralisation, which makes direct control more difficult. As Perrow underlines, "increasingly privatised and deregulated systems […] can evade scrutiny and accountability" [7, p. 379]. Formalising, at least partially, expectations, roles and practices through contracts and rules is thus necessary. Yet, it is of major importance that these documents are aligned with the reality of the work that is being outsourced. This brings forward key questions that involve different actors and concern different steps in the outsourcing life cycle. The first series of questions relate to the very beginning of the life cycle, the decision to contract out: do the decision-makers have a precise knowledge of the activity being outsourced? How has this activity been described in tenders? Does it correspond to reality? The second series of questions can be asked once an activity is outsourced: how are contingencies, unexpected events managed? Do field actors know to who they should report anomalies once they are discovered? If contracts and rules appear ill-adapted to the activity to be performed, can contractors quickly report the problem and modify the rules/contracts?

The second issue relates to coordination and the management of boundaries and interstices. As we have seen, outsourcing increases "the volume of interactions that the system is forced to monitor", with the danger that it "exceeds the organisational capacity to anticipate or comprehend" [7, p. 363]. The chapters revealed key boundaries that need to be diligently managed when it comes to ensuring long-term safety between organisations: some are inter-organisational (between contractors and the principal or between (sub)contractors) but intra-organisational boundaries must be considered too. Among the latter, the boundary between the "elites" [7] (top managers, decision-makers) and field workers is particularly critical. Empirical work reported in this book has described well the difficulties that may arise from the very beginning of outsourcing and throughout its life cycle when organisational elites make decisions about contracting with little knowledge of the reality of the activity, its constraints and the resources (including knowledge) that it requires. In particular, this may complicate the definition of the scope of action of each actor and lead to tensions or conflicts. This relates to the importance of being able to define and then constantly redefine the lines of responsibility and accountability between organisations, especially when it comes to safety. This includes regulators, whose responsibility and accountability has also to be defined. In the case of outsourcing, the issue of coordination seems particularly sensitive to the question of power, and balancing power between organisations and between the elites and those who really operate the system is even more important. Key questions include: who has the power to impose goals on whom? Is the coordination process supported by institutional and socio-material devices? Do mechanisms of coordination and negotiation exist, and do they involve all concerned actors? Is the organisation able to identify the key boundaries that can impact safety performance and how they might evolve over time?

The third issue has to do with knowledge, skills and learning. Inter-organisational boundaries, added to transience, requires re-thinking and re-engineering of learning programmes and practices. When part of the work is outsourced, knowledge issues are particularly critical for four reasons: (1) the necessary skills to perform an activity safely may be unclear: (2) part of the knowledge may be tacit; (3) the contractor or the client may be reluctant to transfer part of the knowledge when it represents a commercial advantage; and (4) the temporary nature of organising may reinforce phenomena of loss of knowledge or organisational forgetting. It is thus important that all organisations along the chain reflect on the following: Are the skills and knowledge necessary to perform the work precisely defined? Do the principal and the clients participate in this definition? What learning devices are implemented to enable knowledge transmission and learning? Are these devices adapted to the nature of knowledge and skills to be learned and transferred? What incentives exist to facilitate knowledge transfer while protecting each organisation's interests?

Finally, we hope that this volume and its conclusions will pave the way to future research that could usefully complement existing literature in safety by further considering the effects of new, yet nearly ubiquitous, forms of organizing that accompanies outsourcing practices. This work does not intend to reduce the complexity of current work situations and to prescribe a "one best way" but rather to acknowledge complexity and manage it better thanks to reflexive practices. The key features and the associated questions, which take the form of an analysis grid, intend to go some way to help all those interested in reflecting on and managing the effects of contracting on safety. It invites researchers and practitioners to engage in further research and fruitful discussions so as to develop innovative practices to manage the outsourcing–safety nexus in new ways.

References

1. M. Bourrier, Safety culture and models: "regime change", in *Safety Cultures, Safety Models: Taking Stock and Moving Forward* (Springer, Cham, 2018), p. 105–120
2. T. Braun, J. Lampel, *Tensions and Paradoxes in Temporary Organizing (Research in the Sociology of Organizations)*, Vol. 67 (Emerald Publishing, Bingley, 2020)
3. N. Brunsson, *The Organization of Hypocrisy: Talk, Decisions and Actions in Organizations* (John Wiley & Sons, Chichester, 1989)
4. S. Gherardi, A practice-based approach to safety as an emergent competence, in *Beyond Safety Training: Embedding Safety in Professional Skills* (Springer, Cham, 2018), p. 11–21
5. A. Hopkins, *Organising for Safety: How Structure Creates Culture* (CCH, Sydney, 2019)
6. R.A. Lundin, A. Söderholm, A theory of the temporary organization. Scand. J. Manag. **11**(4), 437–455 (1995)
7. C. Perrow, *Normal Accidents: Living with High-Risk Technologies* (Princetown University Press, Princetown, 1999)
8. J. Rasmussen, Risk management in a dynamic society: a modelling problem. Saf. Sci. **27**(2/3), 183–213 (1997)
9. B.S. Romzek, M.J. Dubnick, Accountability in the public sector: lessons from the challenger tragedy. Public Adm. Rev. **47**(3), 227–238 (1987)

10. S.L. Star, J.R. Griesemer, Institutional ecology, 'translations' and boundary objects: amateurs and professionals in Berkeley's museum of vertibrate zoology. Soc. Stud. Sci. **19**(3), 387–420 (1989)

Printed in the United States
by Baker & Taylor Publisher Services